W0254340

Choosing and using 4 bit microcontrollers

Choosing and using 4 bit microcontrollers

PHILIP McDOWELL

Marcel Dekker, Inc. New York • Basel • Hong Kong

Marcel Dekker, Inc. offers discounts on this book when ordered in bulk quantities.
For more information, write to Special Sales/Professional
Marketing at the address below

Marcel Dekker, Inc.
270 Madison Avenue,
New York, NY 10016

Published in the United Kingdom by Woodhead Publishing Ltd.
Abington Hall, Abington,
Cambridge CB1 6AH, England

First published 1993

Library of Congress Cataloging-in-Publication Data
McDowell, Philip
Choosing and using 4 bit microcontrollers/Philip McDowell.
p. cm.
Includes index.
ISBN 0-8247-9153-3
1. Programmable controllers. I. Title. II. Title: Choosing and using four bit microcontrollers.
TJ223.P76M38 1993
004.16——dc20 93-15660
CIP

Designed by Geoff Green (text).
Typeset by BookEns Limited, Baldock, Herts.
Printed by St Edmundsbury Press, Bury St Edmunds, Suffolk, UK.

Contents

Preface

'Choosing and using 4 bit microcontrollers' is a book intended for those who wish to do just that - choose and use these particular types of microcontroller. The 4 bit microcontroller has been around for quite a few years, and found its way into a great many products. Many say it is now on the decline, with 8 bit parts becoming cheaper and applications becoming more complex. However, the truth is quite different, with designers of 4 bit microcontrollers fighting hard to stay ahead. As CMOS technology has matured, and integration densities increase with shrinking feature sizes, the 4 bit microcontroller is in the forefront of the integrated peripherals business. Not only does the 4 bit microcontroller offer a complete computer, with internal central processing unit (CPU), program memory, and data store; it also offers many facilities like display drive, analogue to digital (A/D) and digital to analogue (D/A) conversion, and asynchronous serial communications, all integrated in a single device. By integrating many of the system building blocks which otherwise would have to be in separate integrated circuits (ICs), or even built up using discrete components, the 4 bit microcontroller allows designers to achieve cheaper, more compact, less power consuming products.

This book provides information on what 4 bit microcontrollers are available on the market from which manufacturers, giving some indication of where the strengths of each manufacturer lie. Tables of devices, arranged in categories according to a variety of operating parameters, give the reader a head start when delving for the first time into the world of the 4 bit microcontroller, and for seasoned users, give a wider view of what is available than they had possibly seen before.

There are also sections on how 4 bit microcontrollers are used - what type of products and systems they are suitable for and how various commonly required system features can be implemented, considering both hardware and software aspects of the design process. The book is not intended to be a rigorous treatment of the subject, more a source of ideas, with some clues as to the pitfalls awaiting the unwary.

It is hoped that the text will be useful to anyone, student or professional engineer, who has a system to design which needs an embedded controller. It is intended to give an introduction to new users, as well as providing a reference source for those more familiar with the subject.

Introduction

With well over 400 different 4 bit microcontrollers on the market from various manufacturers, the task of selecting the device which will best meet the needs of a given system can be quite bewildering. This task is often further complicated by the fact that the system specification is never fixed, and the capabilities of the microcontroller chosen for the application will often affect the specification. Sales departments are always looking for extra features which can be added to their range of products at no extra cost in order to give them an edge over the competition, and it usually falls to the hapless design engineer to (a) come up with the idea, and (b) engineer it in for free. While perusing the data sheets of the various manufacturers' microcontrollers ideas for extra 'goodies' are often spawned, but there is always the danger that the specification/design will run away with itself, and exceed target costs by a wide margin. It is the aim of this book to help the design engineer to bring the process of selecting the right microcontroller under some sort of control, and prepare the way for good, cost effective design.

The application areas for 4 bit microcontrollers are many and varied, spread over a wide range of manufactured products including aircraft, automobiles, domestic appliances, office equipment, medical equipment, security systems, telecommunications devices, toys, games, and industrial applications such as instrumentation, data logging and production control. It is because the range is so wide that there are so many parts available on the market. The 4 bit microcontroller is not so much a general purpose microcontroller as an application specific standard product (ASSP). Most manufacturers have a small number of core processor units to which they have added a variety of combinations of program storage, data store, input/output (I/O) and on-chip peripherals to meet different application requirements. If a given application is close to one aimed at by a manufacturer then the microcontroller designed for that purpose may well be the most cost effective for the job. However, if the application is a little unusual, not in the general run of the mill volume applications for which these parts have been designed, it may be wise to ignore some of the manufacturers' application labels and examine the basic requirements. Many of the manufacturers of 4 bit microcontrollers are closely allied with product manufacturers, and the existence of a particular part in their range of microcontrollers very often has its origins in a particular product, or range of products made by the company or its associate. A good number of the 4 bit microcontrollers, though, do carry the label 'general purpose', allowing the system designer to make an unprejudiced choice from among the facilities they offer.

In more complex pieces of equipment it may be necessary to use more than one microcontroller, and the skill here is to partition the system suitably so that the specialities of the various types of device can be used to best advantage. For example, one unit may include an operator interface, a real time clock and motor control. The operator interface will be required to drive a display, perhaps a vacuum fluorescent display using a high voltage drive, and also interface with a keypad. The real time clock may be required to remain in operation even when the power is removed from the unit, by using a back up battery. Low power consumption is therefore essential. The motor controller may need special peripherals such as digital to analogue conversion. It is unlikely that all these requirements will be found in a single device, but by partitioning the system in the correct way, and using several 4 bit microcontrollers which include the specific circuitry for each area of application, a more cost effective solution can be achieved than by using a single, say, 8 bit controller and special external peripheral circuitry to perform the specialist tasks required.

In any given system there will probably be one or two factors which dominate the choice of microcontroller, such as power consumption limit or display driver required. There will also be factors such as processing power requirements and amount of program store which are more difficult to quantify, but are still important. Choose a processor with too little program store and the specification cannot be implemented in full; choose a processor with too much program store and the design is not cost effective. This problem has also been addressed by manufacturers of 4 bit microcontrollers, as often they provide three or four parts which in other respects are identical, but contain progressively more program space.

As well as considering power consumption, display drivers and amounts of read only memory (ROM) and random access memory

(RAM), consideration has to be given to what other peripheral devices are integrated on the chip, and what packaging is available. All these things, together with the availability of prototyping and short production run versions, such as erasable programmable read only memory (EPROM), one time programmable (OTP) and piggy-back types, are indicated for a wide range of different manufacturers' devices in Chapter 3. The material contained in this chapter will be of assistance in choosing the right part for the application.

Having chosen a 4 bit microcontroller to suit an application, the engineer then has to make it do the job. This requires a combination of skills in both hardware and software design, so it may well be a team effort. The design team has to be complemented by adequate investment in support tools to enable the design and development to be undertaken in an ordered and economical way. Fortunately such design aids are easily obtained at reasonable cost from all the manufacturers, and the availability of such equipment should not be an overriding consideration when choosing a 4 bit microcontroller. Having chosen a part for one design, there are often strong reasons for staying with the same manufacturer for future designs. Familiarity with the design tools, instruction set and particular methods of one manufacturer all form a strong incentive to stay with that type of device. This could, however, be a false economy. Development tools are not costly – they can often be hired, or even borrowed, for a development project, and although each manufacturer's parts have their own peculiarities when it come to the instruction set and internal architecture, no competent engineer should have much difficulty transferring from one part to another. An engineer familiar with one particular 4 bit microcontroller should have little difficulty in quickly mastering the instruction set of a completely different device from another manufacturer, and be able to write efficient code for it.

Equipped with the necessary tools the team can now get to work. To do this, they will have to generate ideas on how to implement the various functions required, and that is where the remainder of this book comes in. It is intended as a source of ideas. It is not a course in hardware and software design, but rather an aid to the design process. Chapters are included which give ideas as to how various functions can be implemented, both in hardware and software. It is hoped that designers will be able to make use of some of the ideas presented here and develop them for their own applications.

No introduction to a book of this kind would be complete without some form of initiation into the particular jargon of the subject. To make sense of the remainder of the text it is important that the reader has the same understanding of the jargon as the author. Often specialist jargon develops differently in different places, so below is a guide to the meanings of terms used in this book. This is not intended as a guide to the specialist language of the world of 4 bit microcontrollers as a whole, but a means of avoiding confusion over what a particular term means when used here.

4 bit microcontrollers are essentially digital devices, and hence work in binary code, with binary digits, or bits. Bits are combined in various wordlengths, and the commonest short wordlengths are related as follows:

2 bits	= 1 crumb	
2 crumbs	= 1 nibble	(4 bits)
2 nibbles	= 1 byte	(8 bits)

The term nibble will be used throughout this text to mean a 4 bit word. A nibble is sometimes confusingly referred to as a digit, since it is often used to store a decimal digit in binary code, but most manufacturers use the term nibble.

Memory space, particularly code (program) store, is usually expressed in ks, 1k = 1024 (2^{10}). This is often in bytes, but sometimes in words of instruction length, which may be a slightly strange number like 14 bits.

Numbers are often expressed to the base 16, rather than the usual ten, in which case they are followed by the letter H, e.g. 1k = 1024 = 400H.

There are a great many abbreviations used in the industry. Below is a list of the ones used in this book, together with their meanings:

ALU	Arithmetic and logic unit
A/D	Analogue to digital converter
BCD	Binary coded decimal
CMOS	Complementary metal oxide silicon
CPU	Central processing unit
D/A	Digital to analogue converter
DIP	Dual in-line package
DTMF	Dual tone multi-frequency, tone codes used for sending numbers over telephone circuits and other voice frequency channels
DTS	Digital tuning system
D2B	Domestic digital bus

EEPROM	Electrically eraseable PROM
EPROM	Eraseable PROM (using ultra-violet light)
FIP	Fluorescent indicator panel
IDC	Image display controller
I/O	Input/output
LCD	Liquid crystal display
LED	Light emitting diode
lsn	Least (or less) significant nibble
NMOS	N-channel metal oxide silicon
msn	Most (or more) significant nibble
OTP	One time programmable
PABX	Private automatic branch exchange
PGA	Pin grid array
PIR	Passive infra red detector (for security systems)
PLC	Programmable logic controller
PLCC	Plastic leadless chip carrier
PLL	Phase locked loop
PMOS	P-channel metal oxide silicon
PROM	Programmable ROM
PWM	Pulse width modulation
QFP	Quad flat pack
RAM	Random access memory (read/write)
ROM	Read only memory
SDIP	Small dual in-line package
SLR	Single lens reflex (camera)
SOIC	Small outline integrated circuit
SOP	Small outline package
TTL	Transistor-transistor logic
VCR	Video cassette recorder
VFD	Vacuum fluorescent display
VFT	Vacuum fluorescent tube
ZTAT	Zero turn around time (Hitachi OTP and EPROM)

1 A brief history of digital computing

The history of digital computing can be traced back to the nineteenth century, but it was not until the 1930s and '40s that progress really began to be made. The military requirements of World War II had a major influence, making resources available for development work which otherwise would have taken place much more slowly. A major obstacle in the way of progress though, was the thermionic valve. Whereas the valve was of great value in developments like radio and television, it was a problem in computing because of the large numbers of valves needed to perform quite basic functions. One of the main problems was the reliability, or lack of it, of the valves themselves. There were also problems of space and cooling, as valves were comparatively large devices, and relied on being heated to operate. Another major early problem was that of memory, where the ingenuity of scientists and designers was stretched to the limit in devising techniques for providing the memory that was essential for computers to operate. However, a press release by Bell Labs on 1 July 1948 changed the world as far as electronic computing was concerned. With the advent of the transistor which it announced, the computer really began to come into its own, and institutions with sufficient funds began to invest in them in a way that they had been reluctant to in the days of the valve - government, universities, banking and finance institutions were all quick to see the advantages to be gained from the use of computers, and also had the resources to invest in them. Computers were, in these early days, very expensive pieces of equipment, involving a great deal of manual labour in their construction. Then came the era of the integrated circuit, and computers became at the same time smaller, cheaper and more powerful. The result, of course, was a great blossoming of applications. Computers began to be used where no one had believed possible a few years earlier. The emphasis, though, was still on data processing, and the special requirements it imposed, such as the desirability of long wordlengths.

'Mini computers' started to be made with 12 and 16 bit words, but most 'serious' computing was performed on machines with 24, 32 or 64 bit words, the aim being to perform high precision floating point arithmetic as quickly and as efficiently as possible.

When the first 'Computer on a chip' was unveiled, 20 or so years ago, it was a 4 bit machine. It was for its time a major step forward in terms of integration, but the reaction in many quarters was 'What on earth can you do with only 4 bits?'. The answer surprised many people, as quite a lot could be done with only four bits. That device, of course, relied entirely on external memory, and the memory bus was only 4 bits wide, allowing only 4 bits of instruction to be fetched at a time. Needless to say, at least two fetches were required per instruction as the internal instruction register was 8 bits long. Since those days great strides have been made, advancing the capabilities in both design and processing, and by today's standards that early pioneer seems extremely primitive. However, the idea of working with a 4 bit data word has never lost its attraction in particular applications.

Pocket calculators

The advances in electronics, particularly digital electronics and computing began to find applications outside the mainstream computing developments. Computers were getting more powerful all the time as the technology developed, and new architectures were designed to improve their performance, but engineers were also applying the techniques in other ways, and the pocket calculator is a prime example of this. The pocket calculator is a very visible product of the advances in electronic technology over the past few decades. When such equipment began to appear it soon caught the public's imagination, and everyone wanted one. This huge market potential invited corresponding amounts of investment, and allowed a great deal of development effort to be put into this area.

The vast majority of calculators work with decimal numbers, and internally in binary coded decimal (BCD), where each decimal digit is represented by a 4 bit binary field (referred to in this text as a nibble). Internally it made sense to stick to BCD for all the arithmetic operations, to avoid the necessity of converting all the input to true binary, and then converting it back again to display it. With as big a market as the calculator market to aim at, there were strong incentives for developing lower cost, lower power, smaller calculators which the general public were demanding. To achieve this it was necessary to reduce the number of components to a minimum, and integrate as many as possible into the main calculating chip. In parallel with these developments in integrated circuit technology, new devices were becoming available for displays and for the power source. These developments had their impact on the chip design as well, as chips had to cope with driving liquid crystal displays (LCDs), and operate from solar cells, mercury cells, lithium cells, and so on. Much of this work has its parallels in the development of 4 bit microcontrollers, and has probably contributed to their development to no small degree.

Digital watches

Another product which has become almost universal in its appeal and application is the digital watch. The ability to produce such an accurate timepiece at such low costs is taken for granted by people today, but not many years ago it would have seemed like an impossible dream. The development of the digital watch, like the calculator, has attracted a great deal of investment because of the huge markets involved, and the potential for profit. Like the calculator, the operations to be performed within the watch chip lend themselves to 4 bit operations, and making special purpose watches is one application area for 4 bit microcontrollers today. Looking at the list of companies who manufacture 4 bit microcontrollers it will be seen that some are strong in the watch business, and their involvement in 4 bit microcontrollers has no doubt sprung from their development work in digital watches.

Microprocessors

The 'Computer on a chip', or microprocessor as it became known, soon developed to an 8 bit device, and later to 16 and 32 bits. Engineers working on the design of logic systems had become used to racks full of printed circuit boards, each board containing rows of chips each performing a fairly simple logic operation. They now saw the possibility of integrating vast areas of circuitry into a convenient, single, programmable chip, saving not only on design time, but also on component count, circuit board area, power supply capacity, and also the total space for the system. It was now possible to build in more flexibility than ever before. Systems could be designed with built in diagnostic programs to aid the service engineer. All sorts of possibilities opened up, and microprocessors began to be used in a whole new range of applications, particularly in the field of 'embedded control'. Any device or system which had in the past used some sort of digital logic, from relays to transistor-transistor logic (TTL), now lay open for the application of the microprocessor.

Memory technology

In parallel with microprocessor development was the important work on memories. Looking back it is amazing to see the ingenuity that was put into developing non-semiconductor memories, and the degree of sophistication achieved, but it was not until the semiconductor memory became a viable proposition that the truly integrated computer became possible. The reduction in costs, and consequent increase in the application of computers in general was increasingly opening up the market for memory devices. The computer designers and software engineers seem to have an insatiable appetite for memory, always looking for more, and always developing applications which used what was available to its limits. Again, a big market invited big investment, and as a result a great deal of engineering ingenuity has, over the years, been invested in memory technology. It is this development of the semiconductor store, with its demand for ever increasing densities, that has driven the developments in semiconductor processing technology, pushing down feature size, pushing up chip densities, chip sizes and wafer sizes. This in turn has made possible the advances in integration that we see today, and it is the integration of memory with processing logic that has given birth to the family of devices known as microcomputers and microcontrollers.

Traditionally, the computer, whatever its size, consists of a central processing unit, memory, and input/output (I/O). The development from microprocessor to microcomputer mainly consisted of integrating the memory with the processor, leaving the I/O to external components. The memory was divided into two parts, read only memory (ROM) for storing the program, and random access memory (RAM) for storing run time variables, and data read in, the results calculated, etc. Also integrated was the system clock, allowing the user to employ just two or three inexpensive components to determine the clock frequency, rather than having to supply a complete oscillator circuit. The microcomputer has been a great success, and there are countless systems operating under the control of such devices, particularly 8 bit parts.

The early pioneers

The majority of the early progress in microcomputing was made in America, with companies like Intel, Rockwell, Texas Instruments and National Semiconductor. Intel were pioneers in the field with their MCS-4/40 microprocessor system. This was a family of parts including a microprocessor and a range of peripheral chips from which it was possible to build a processing system. The microprocessor, the 4004, was a P-channel metal oxide silicon (PMOS) device with a 4 bit data bus, 8 bit instruction word, and a 12 bit program counter, allowing access to 4k bytes of program code. The internal architecture was designed with a stack allowing 3 levels of subroutine nesting, and a scratchpad giving 16 nibbles of internal RAM. The instruction set was limited to 46 instructions, including conditional branches, the use of subroutines, and indirect memory access, as well as binary and decimal arithmetic modes.

The 4004 was later joined in the range by the 4040, an enhanced version with 60 instructions, adding logical operations, and the ability to read program memory. Subroutine nesting was increased to 7 levels in this enhanced device, and the scratchpad to 24 nibbles. An interrupt mechanism was also introduced. Both parts required a power supply of 15 volts, specified as +5V and −10V, and had an instruction cycle time of about 10 microseconds. To accompany the microprocessor it was necessary to add a clock generator, a memory interface chip, which separated the data and the address, the program memory (ROM), a RAM, and whatever I/O was required by the system. There was a variety of I/O devices available, such as a keyboard and display interface, and several general purpose I/O expanders. I/O was included with memory in several devices. The MCS-4/40 was available in the catalogues up until the late 70s, but by the early 80s had dropped out of the range. Intel had by then committed itself to longer word-lengths, and had introduced their 'single component' microcomputer, the MCS 48 range, the forerunner of what has become a de facto standard in 8 bit embedded control, the 8051.

Rockwell International was another American company in the forefront of developments in 4 bit microcomputing. Rockwell designed their first programmable CPU as early as 1967, and by the mid-70s claimed to have supplied over a million 4 bit microprocessor systems. Their PPS-4 family aimed at providing a two chip solution, one chip providing the CPU functions, with built-in clock generator, the other providing ROM, RAM and I/O, with up to 2048 × 8 bits of

ROM, 128 × 4 bits of RAM and 16 individual I/O lines. Peripheral chips were also available for a variety of other purposes, including interfacing to more read/write memory, both semiconductor and non-semiconductor types. The CPU included such features as binary and decimal arithmetic, logical operations, conditional and unconditional jumps, subroutine calls (with 'unlimited' nesting) and an interrupt system. Computing speed was such that two decimal numbers could be added together at the rate of 30 microseconds per digit. Processing was still in PMOS, with a 15 volt power supply requirement, but N-channel metal oxide silicon (NMOS) was now established as the primary memory process, and by the mid-70s they had reached the size of 4k bits per chip. Program ROM in these systems was of course mask programmed, that is the contents of the memory were fixed by mask patterns used in the manufacturing process. So as well as developing these 2 chip 4 bit processing systems, Rockwell was offering ROM/RAM/I/O chips which were unmasked, for use in prototyping. They also had a fairly comprehensive range of program development aids, including their 'Assemulator' modules where program could be held in read/write memory, or in PROM, and providing interfaces to such peripherals as teletypes and tape readers. Programs were written in assembly language, and entered as data to a cross assembler which was written in Fortran, a language then almost universally used in scientific and engineering circles, and available on many in-house mainframe computing facilities, and also from a number of timeshare facilities, where users were linked by telephone and modem to a central computing installation. Libraries of programs were also available, giving useful subroutines, initialisation procedures, and programming examples.

In 1976 Rockwell announced their PPS-4/1, a one chip microcomputer with 1344 × 8 bits of ROM, 96 × 4 bits of RAM, and 32 I/O lines. This part was an extension of the PPS-4 system, and as well as introducing single chip computing, began adding integrated facilities such as a serial I/O capability. By the end of the '70s they had extended the range with the introduction of complementary metal oxide silicon (CMOS) versions, operating at lower supply voltages, and lower power levels. The M78LA device included 2048 × 8 ROM, 128 × 4 RAM, 35 I/O lines, and included display control, and speaker drive. It operated from as low as 6.5 volts, with a typical 15 milliwatt power consumption, in contrast with the typical 75 mW power required by the PMOS parts. The list of applications for the PPS-4/1 reads like that of one of today's microcontrollers, including such things as domestic appliances, automobile features, cash registers, coin changers, environmental control, dot-matrix printers, keyboard/ display controllers, hand held and arcade games, petrol pumps, traffic controls, instruments, language translators, machine tools, process controls, radio tuners and scanners, consumer and commercial scales, security systems, desk top calculators, telephone auto diallers, switching and auto answering equipment, timers, clocks, TV tuners, etc. The 4 bit microcontroller was by then well established as a universal work-horse in embedded control, but developments in 8 and 16 bit devices were coming on apace, and the American manufacturers, with their eyes on the high spenders such as the military and space exploration programmes, found that the profit was going out of the 4 bit market, and began to desert it, leaving the Japanese electronics companies, with their emphasis on consumer goods, to make the running.

Texas Instruments were great rivals with Rockwell, and by the early '80s they claimed 'Currently there are over 600 TMS 1000 programmations or 60 million units installed in the field, making the TMS 1000 the most pervasive microcomputer in the world'. The TMS 1000 family was a range of parts with program ROM from 512 × 8 bits up to 4096 × 8 bits and from 32 to 128 × 4 bits of scratchpad RAM, and a variety of I/O options. There were parts available in 9 and 15 volt PMOS and in 5 volt CMOS, the CMOS parts being of a slightly more advanced design, offering a more flexible subroutine facility. An interesting feature available with the TMS 1000 family was the integrated display drive facility whereby some of the output lines could be coded up through a section of PLA on chip to provide, for instance, seven segment encoding from a numeric internal value. Some parts also had high voltage (35 V) capability, allowing them to drive vacuum fluorescent displays. Texas Instruments also provided a comprehensive range of program development tools, with system evaluator chips that operate from external EPROM, and complete stand alone development systems for program development. Some parts were available ready programmed for specific tasks such as a microwave oven controller, and a standard timer device suitable for use in clocks, time switches, etc. Texas Instruments, like most other American semiconductor companies, has moved on to more advanced devices with larger wordlengths and more processing power, but they have not abandoned the 4 bit field entirely. Although the TMS 1000 family has now dropped out of the product lists, there is still the TSS 400 family, described as 'sensor signal processors', and the TCM

8305/6/7/8 family, which is aimed at the telephone market, designed for application in telephone sets.

Another American company, National Semiconductor, is also worthy of mention in this brief historical review. National is the only other American company still active in the 4 bit microcontroller field, with their COP 400 family of devices. This family of parts, which began development in the '70s, was well established by the end of that decade, and offered some fairly advanced features such as low voltage operation, down to 2.4 volts for some parts, and sub-milliamp supply currents, as low as 5 microamps in the 'sleep' mode. As this is a range still currently available, more details of its facilities will be found in Chapter 3.

The Americans were really the pioneers in 4 bit microcontroller design, establishing during the '70s the basic architectures which are still used today. The fact that most American companies now show little interest in this area, with new developments only taking place in rather specialised fields, is a reflection of the fact that 4 bit microcontrollers are used mainly in application areas where the Far East has a dominant market position. European companies have never apparently shown very much interest in the field, and the only European company now playing a significant part is SGS-Thomson, who provide a second source for the National Semiconductor range of parts. During the '80s the centre for development in 4 bit microcontrollers moved to the Japanese semiconductor industry, where it has been a picture of continuously applying advances in processing technology to improve the specifications of devices available. The main thrust of this has been to provide parts with more and more integrated peripherals, giving families of devices which share a common core CPU, but have different combinations of integrated peripherals to suit different applications.

The modern 4 bit microcontroller

The modern 4 bit microcontroller is a complete computer in miniature, having all the essential elements that any computer requires built in. As mentioned previously, it contains a central processing unit, program store, data store, and a variety of input and output systems. Very often it also contains special purpose peripherals, such as display drivers, asynchronous communications circuits or analogue to digital and digital to analogue converters. The key to the success of the 4 bit microcontroller is this peripheral integration. The art of choosing a 4 bit microcontroller is to select a device which has the right combination of storage, I/O, and peripherals for the application.

The 4 bit microcontroller is a self-contained unit requiring minimum external circuitry to make it work. Often, all it needs is a power source, an external oscillator component such as a crystal or ceramic resonator, and some decoupling capacitors. The internal architecture of the 4 bit microcontroller often resembles that of bigger microcontrollers and microcomputers. It is called a 4 bit device because it handles data 4 bits at a time. Program memory is not restricted to 4 bit words – program store is always at least 8 bits wide, and often wider, up to 16 bits. Parts with 8 bit program store will have many instructions in their repertoire which are 2 or 3 bytes wide, whereas it is the aim, when increasing the program wordlength, to reduce all, or a large percentage, of the instructions to a single word, and thus minimise instruction fetch time. This is often important with the slower, bottom end devices which, because of their comparatively slow clock rate, cannot afford the extra clocks necessary for multiple byte instructions.

At the top end of the range of parts on the market, architectures closely resemble those of the 8 bit families, having sets of general purpose registers, stacks in RAM for storing return information in subroutines and interrupts, and multimode RAM addressing. Sometimes the registers bear the same names as familiar 8 bit parts, and the programmer has to remind himself or herself from time to time that these are 4 bit registers. At the bottom end, however, the devices are very simple, with limited instruction sets, register sets mapped into part of RAM, and no provision for handling subroutines or interrupts. The amounts of RAM and program store also differ widely. RAM in the bottom end devices may be as little as 16 nibbles, but at the top end some parts have over 1k nibbles of general purpose RAM. In addition there are some special purpose parts available with up to 4k nibbles of RAM. Program memory, which may be mask programmed ROM, or field programmed EPROM, varies from 512 up to 32k words, or even more in new parts under development.

The designer faced with the problem of making the right choice of microcontroller for a new system does not have an easy task. There are so many parts to choose from, and so many design aspects to be considered that it is extremely difficult to be sure that the best choice has been made. It should not be too difficult to draw up a short list using the information contained in the following chapters, but that list may include devices which are quite different in terms of internal architecture and instruction sets. It is important to establish that the performance required by the system can be achieved, and so it may be necessary to code up critical sections for competing parts to get a feel for how easy it is going to be to achieve the required goals. Sometimes there will be a choice of parts from the same family, perhaps offering a range of operating voltages with corresponding power consumption and operating speeds.

Further reading

In any book of this nature, where subjects are tackled in a general way, the reader will probably need to look at some aspects in more depth. Usually there will be a list of books to refer to, and the reader selects books from the list which look as if they will provide the necessary in-depth information. With this book, however, there is no such list. The reason for this is that the majority of information presented here has been drawn from manufacturers' data books, and it is these data books which really give the detailed information required on any particular device. When a short list of possible devices for a particular project has been drawn up, it is essential to obtain as much information as possible from the manufacturers of those products, and most manufacturers supply such information in the form of data books covering a range of devices. Manufacturers also publish application notes, software handbooks and other guides to the use of their products. It is always worthwhile asking for a complete list of publications produced by the manufacturer so that any which might have useful relevant information can be obtained.

There are a great many text books on many of the subjects covered by this book. Many books have been written to help students and designers with the tasks of developing circuits based on microprocessors and microcontrollers. All aspects of hardware and software design have been covered, and the reader will have little trouble finding a range of such books on the shelves of libraries and bookshops. Many of these books may appear at first sight to be working towards applications too far removed from 4 bit microcontrollers. Very often a text book will concentrate on applications for a particular processor, probably working with 8 or 16 bit data wordlengths. However the general principles involved still apply to the 4 bit world, and many of the techniques introduced and software algorithms published can simply be adapted to operate on a 4 bit machine, even when its architecture is very different from that of the original target machine. The reader is therefore urged to delve into a rich source of ideas. It is hoped that the introduction to the 4 bit world provided by this text will serve as an appetiser, and encourage the reader to explore more widely.

2 Applications

4 bit microcontrollers are used in such a wide variety of application areas that this chapter stands the risk of becoming a list of all areas of the application of electronics in general. At one end of the spectrum, because of their low cost, they can be used in areas where, traditionally, no electronic control has been applied, and they replace electro-mechanical controllers, or perhaps introduce a measure of control where there has been none before. In this way they enhance the functions of a product or system, and hopefully make it easier to use. At the other end of the spectrum of processing power there are a number of application areas where 4 bit devices will be rivalling 8 bit parts, and the designer of a new system will have to examine the competing specifications with care to make sure that a correct choice has been made.

To make it easier to understand the diversity of the application areas for 4 bit microcontrollers they have been divided here into a number of broad classes. The groupings below are not intended to be an exhaustive list of all possible application areas, but to give some idea of the capability of 4 bit microcontrollers, and of their versatility.

Low power battery operated devices

This group includes a wide variety of devices, including special purpose wrist watches, pocket calculators, miniature games, electronic diaries, etc. They all share a number of features, for example they all usually have a liquid crystal display, they are required to be as small as possible and consume virtually no power. Most of the devices in this class have a time function built in, so require an accurate oscillator, usually employing a 32 768 Hz crystal. The semiconductor industry has responded to the demand for this type of equipment by producing a variety of parts suitable for such use. Parts are available which will operate from 1.5 or 3.0 volts, to suit the mercury and lithium cells that are most appropriate here.

In-built LCD drivers are common features of these chips, usually not more than 4-way multiplexed, a common size being 32 × 4, driving a maximum of 128 display elements. Parts are available to drive more segments; OKI make a device which drives 58 × 4 (total 232) segments, and Toshiba have a part which has 10-way multiplexing, and drives 44 × 10 display elements. Other integrated functions which are useful in this type of application include bleepers, or melody generators, dual tone multi-frequency (DTMF) tone generators, and some parts have the capability of handling analogue signals allowing, for example, temperature indication. Packaging is an important consideration in this group, and as a result devices are usually available in flat packs for surface mount and often as bare chips.

There is another class of products closely related to those above which have no integral power source, but will pick up power for example from an inductive loop. Such items as smart cards and transponders operate in this way, and it is clearly important that power consumption is kept to a minimum. Often there is no display on the unit. All information is communicated through the interrogating system via the inductive link. In this type of application, all memory, including read/write memory, has to be of a non-volatile type.

Some battery operated applications are not so power sensitive, but other features such as low cost become important. In the toy market, simple motor control functions may be required, or perhaps the ability to operate a speech synthesis system. Here the power used by the microcontroller is small compared with other parts of the unit, but the control function is quite limited and the 4 bit microcontroller supplies a cost effective solution.

White goods and domestic controllers

4 bit microcontrollers are used extensively in many domestic appliances such as cookers, microwave ovens, deep fat fryers, dishwashers, sewing machines, washing machines, tumble driers, central heating controllers, air conditioning controllers, etc. They have often replaced mechanical timers and sequencers, and have at the same time increased functionality by their flexibility of application. They are also increasingly being used to add functionality to devices which traditionally contained no control beyond perhaps a simple thermal switch. Such things as electric irons, toasters, coffee machines, kitchen scales and even shavers will contain 4 bit microcontrollers. The computing load is usually very small, and the 4 bit controller is much more cost effective

than the more powerful 8 bit device because of its lower cost and additional in-built peripheral functions.

Most of the appliances in this category are powered from the mains electricity supply, so power consumption is not usually critical. There is usually plenty of space available and so the availability of miniature packaging is less important. The types of features offered by the manufacturers which come to the fore in this area are vacuum fluorescent display (VFD) drivers, output ports with open drain circuits allowing them to be pulled up to a higher voltage than the controller operating voltage, analogue to digital conversion, pulse width modulation (PWM) outputs for controlling motors or other devices requiring an analogue signal, and high current capability outputs (for driving light emitting diodes (LEDs) for example). A buzzer drive is often a useful feature, as is a watch-dog timer. The watch-dog will reset the controller if the internal program fails for some reason. In applications where there is likely to be electrical noise this can sometimes cause corruption of internal registers causing the program to operate incorrectly. The watch-dog timer then acts as a safety device to put the controller back into a known state. Designers should be cautious however about relying on this as the sole protection against malfunction. Wherever there is a risk of serious consequences of malfunction a microcontroller must always be backed up by some extra external safety device. For example, this can often be a simple thermal cutout which will disconnect an appliance from the mains electricity supply should it be in danger of catching fire.

Brown goods

Modern homes now bristle with electronic appliances to entertain and keep us informed, e.g. television sets with teletext service, video cassette recorders (VCRs), hi-fi systems, satellite decoders, camcorders, etc, each with its own remote controller. Each of these items will probably contain at least one 4 bit microcontroller, and some will contain several. Because this sector of the electronics industry is dominated by the Japanese manufacturers, and because the majority of manufacturers of 4 bit microcontrollers are also Japanese, there is a strong tie-up between the component design and equipment design. As a result there are many 4 bit microcontrollers with features specially suited to these applications. A VCR will typically contain four of them, one to handle the control panel and display, one to handle timing, and the timer recording functions, one to control the mechanical functions such as motor drives and cassette eject mechanism, and one as the remote control receiver. Digital tuning, referred to by some manufacturers as DTS (digital tuning system), is another function served by 4 bit microcontrollers which is often employed in VCRs and TV sets. There are devices from several manufacturers which contain phase locked loop (PLL) circuits for this purpose, and can be employed in audio equipment as well as video. Another special function available for the video scene is the integrated on-screen display controller/driver, sometimes referred to as an image display controller (IDC) for TV and VCR, allowing channel numbers and the like to be displayed on the TV screen. Some of these have the capability of displaying up to about 200 characters.

Inter-device communication is clearly a necessity in systems of this nature, and most of the parts designed for these applications include serial I/O ports. The domestic digital bus (D2B) is now becoming established as a common interface standard for such equipment. This allows a number of appliances to be connected on a 2 wire data bus for the transfer of control data and other information. The bus can support up to 50 devices over a distance of 150 metres at data rates of up to 7760 bytes per second. Such a bus system could for example be used to transfer data between a VCR and the device in the television which provides the on-screen display facility. Implementing such a data bus may require use of a special interface chip, but this will probably have been designed specifically to operate with a 4 bit microcontroller. A data bus of this nature can clearly be used in application areas other than the domestic appliance area. One area where they are to be found in increasing numbers is in automobile electronics, providing communication between various dashboard instruments, trip computer, etc.

Analogue signals also have to be handled, and there are many parts available with either simple comparator inputs or multi-channel A/D converters, and also PWM outputs. The choice of analogue facilities is quite wide, and a designer should have little difficulty in finding a chip with the converters appropriate to a specific system.

For the audio devices, including tuners, CD players and audio cassette machines, the requirements are very similar, and many chips are designed for use in both audio and video systems. DTS, remote control receivers and motor control are all features employed in audio systems, which may also include the clock facilities, control panel and display drive features found in the VCR.

Camcorders present more of a challenge, as they combine many features of the VCR with a video camera, and have the added constraint of being battery powered and as small and lightweight as possible. The power consumption is not as critical as with watches, calculators and the like, as the power consumption in a camcorder is dominated by the requirements of the various motor drives, but a power hungry microcontroller could well be an embarrassment to the designer. As the functions of the camcorder are so complex, they usually use a more powerful microcontroller for overall control. This will probably be a 16 bit device, and so the 4 bit microcontroller is employed as a peripheral to the main controller. Communications then become important and parts designed for this application will often have 8 bit or even 16 bit parallel communication facilities.

Remote controllers are an important feature of this group of products. Infra-red remote control is expected by the consumer to be available on all home entertainment products, except perhaps the camcorder. Some of the more sophisticated single lens reflex (SLR) stills cameras also employ remote control. As a result there are many 4 bit microcontrollers available aimed specifically at the remote control market. Mitsubishi, for example, make a range of parts for infra-red remote control transmitters which have various special features making them suitable for this specific application (and unsuitable for any other). Other manufacturers also produce similar parts. One principal feature of these parts is the provision of a modulated output to drive the infra-red LED which transmits the output control message. Modulation is commonly at 38 kHz, though other modulation frequencies are available. The parts are designed to operate a keypad matrix for input, and commonly provide for the selection of a 'custom code' to prevent confusion of signals, and the wrong piece of equipment responding to a transmitted command. Such custom codes are sometimes allocated by the microcontroller manufacturer, who will keep a register to ensure that all their customers use unique codes. Remote control devices are becoming more and more sophisticated all the time, and manufacturers of microcontrollers are responding with more complex parts. The learning remote controller is an example of this, where provision is made for reception of infra-red signals as well as transmission, so that one controller can be 'taught' the codes of other controllers, allowing the end user to control a multiplicity of devices from a single controller. Other features seen in remote controllers are bar code readers for programming the VCR, built in real time clock and simplified programming of the remote controller to allow the remote control unit to handle the timer record function. As more and more home appliances with infra-red remote control become available the possibilities multiply. Interfacing to the telephone or modem would open up all sorts of extra functions, and when interfaced to a personal computer, an infra-red remote control can supply all sorts of services to the disabled.

One further possibility introduced by the increasing use of microcontrollers in all domestic appliances, whether of the white goods or the brown goods category, is the concept of utility load management. The simplest case is for the electricity company. When a sufficient number of domestic appliances are fitted with suitable controllers, it gives the electricity company an alternative to load shedding in the event of the demand exceeding the available supply, as occasionally happens, for example at times of severe weather conditions, especially if part of the supply network gets cut off. Under these circumstances the electricity company sends an appropriate control code over the power lines, which will be received by the appliances able to respond. These 'intelligent' appliances can then switch into an emergency, power saving mode, and cut electricity demand instantaneously. It may be, if suitable agreements can be reached, that manufacturers will be required to provide such facilities in their appliances when sold into certain countries, or geographical areas, so that such control can be effectively used.

Office equipment

Most of the equipment in the modern office is electronically controlled to some degree. Typewriters, word processors, printers, plotters, calculators and photocopying machines have all undergone revolutionary change in recent years. Much of the revolution in the office is due to more powerful microprocessors than the 4 bit microcontroller, but it does find its place in many parts of the office. Typewriters, for example, commonly have the sort of facilities that a 4 bit microcontroller can provide. Storage of a line or so of type, display on an LCD, font control for a dot matrix type printhead, and communication with other office equipment are typical of such facilities, additional to the essential tasks of keyboard scanning and motor control. Printers have similar requirements and all the functions of communications, motor control, font selection and so on can be handled by a 4 bit microcontroller. Many peripherals to the ubiquitous personal computer are controlled by an embedded 4 bit microcontroller. The availability of such a wide range of controllers simplifies the design of devices such as plotters and scanners, and at the same time making it easy to design them with operator interfaces which make them easy to use.

The office calculator is another prime candidate for the 4 bit microcontroller, often having an LCD display together with a small printer. The microcontroller allows the designer to put in whatever special features might be needed in a particular model, and a range of models can be produced having the same basic facilities but different extra functions, all operating from the same microcontroller where a few inputs are used to determine which options will be made available. The use of the 4 bit microcontroller in the photocopying machine has allowed it to become both more compact and easier to use. Again, the in-built peripherals which provide the means for display driving and motor control, as well as the general purpose I/O used for keyboard scanning and checking the various sensors around the machine, give a compact and cost effective controller which can provide a simple, easy to use, interface to the operator. Obviously the 4 bit microcontroller is not capable of providing the kind of facilities required by large, high volume copiers – their place is at the bottom end of the market where products are sensitive to size and cost.

Other devices found in the office environment which could be controlled by a 4 bit microcontroller include such things as postal scales (which indicate the stamp value required), electronic coin and note checkers where money is being handled, clock/calendar units, electronic signboards, and possibly lighting and environment control.

Audio visual equipment

More and more sophistication is expected nowadays from audio visual presentations in applications ranging from sales presentations to permanent displays in museums. Co-ordinated multi-media presentations, combining input from a variety of media such as multiple slide projectors, video tape and disc, audio tape and disc, and possibly film as well, present a challenge, especially when they have to run unattended at an exhibition, or are required to be operated by people with no technical skills. When analysed, the control functions required, such as lamp brightness, sound volume and operating switches, are just the types of function that a 4 bit microcontroller can provide, and the computing power required is often quite modest. Thus, they would seem to be the obvious choice for providing such control.

Telecommunications

The liberalisation of telecommunications legislation in the UK has led to an explosion of new products coming on to the market, and with the growth in mobile and personal communication facilities which is now under way, there promises to be a continuing market for new telecommunications equipment of one sort or another. In the office are to be found small PABXs, feature phones, cordless phones, answering machines, facsimile and telex terminals, and modems for computer communication. Out of the office are mobile phones of various sorts and paging devices of increasing functionality. All these can and do employ 4 bit microcontrollers in one way or another. The manufacturers have addressed this valuable market, and provide facilities such as DTMF diallers and receivers, large memories for repertory dialling, low power consumption and extra thin packaging, as space is often at a premium, especially in the portable devices.

Peripheral functions are integrated to a much greater degree on 4 bit microcontrollers than with any other group of controllers, but in telecommunications systems there are many functions required that are not available as integrated functions. Such things as analogue speech circuits and radio circuits for cordless phone, cellular phones and other portables, are not available as integrated peripherals, but often manufacturers make separate circuits to perform these functions which are designed to interface with the appropriate microcontroller.

Automobile applications

The opportunities for the application of electronic control in the automobile is another area where rapid growth has been seen in recent years, and it is a very important sector of the industry because of the volume of parts involved. Many of the areas where electronic control is being introduced are beyond the capability of the 4 bit microcontroller. Such aspects as ignition and fuel injection control, power steering, suspension and anti-lock braking systems are well beyond the scope of the 4 bit part because of the complexity of the computation required and the speed at which it has to be carried out. However the 4 bit microcontroller will find a place in the dashboard, the trip computer, the car radio, the car telephone, the central locking and alarm system, and also in making safety checks – for example, that all occupants have their seat belts properly fastened. One interesting application of the 4 bit microcontroller in the automobile is to allow the electrical wiring of the vehicle to be made much simpler. Traditionally, each item of electrical equipment has its own individual wire from the battery or the fuse box, via the dashboard switch. This approach was quite satisfactory in the early days of motoring, when there were very few electrical parts on the vehicle, but as the number of functions requiring electrical power has increased, so has the complexity and cost of the wiring loom. Not only does this increase the manufacturing cost of the vehicle, but it also creates reliability and servicing problems. By using 4 bit microcontrollers it is possible to design a multiplexed wiring system, where each electrical device is supplied with power from a 'ring main' type wiring system with an intelligent switch (the 4 bit microcontroller), on a simple control bus. This allows a much simpler wiring arrangement to be used, and also gives the opportunity for extra functions to be added, such as detection of failed bulbs and overload protection. It is not just cheaper, it is better too.

The environment in an automobile, particularly in the engine compartment, is a difficult and challenging one for electronic equipment. It requires the use of parts which can operate in a wider range of temperatures than normal, and it makes demands on the designer, making sure that a design is robust and reliable enough to endure the levels of dirt, vibration and electrical interference it is likely to be subjected to, and that all safety considerations have been accounted for. This applies to both hardware and software aspects of the design, and special techniques may have to be applied in the design process, such as Formal Methods, in order to satisfy the regulatory requirements.

It is not just the semiconductor companies who are active in this field – the automobile manufacturers themselves now have research teams working in the field of semiconductor devices, looking at aspects which are of particular importance to their industry. Toyota, for example, have reported work on a high reliability 4 bit microcontroller, which has various special features making it particularly reliable. They have designed on the basis of a minimal instruction set, and included memory parity checking, and the ability to provide default outputs in the event of failure.

Security systems

4 bit microcontrollers are the obvious choice for providing 'intelligence' in small alarm systems for homes, offices, shops and other premises where the number of alarm sensors is relatively modest. They are also useful in entry control systems, providing controlled access to secure

areas, and communicating with a large alarm system if necessary. In such systems they can provide display driving, keyboard scanning, interface with card readers, scan inputs from sensors such as passive infra-red (PIR) detectors, window security foil, proximity switches, vibration detectors, etc, as well as controlling the drive to alarm sounders, and interfacing to the telephone system. At the very simplest end they provide a cost effective means of controlling single zone intruder alarms, where there might be a single sensor, a keypad and an alarm sounder. Low power microcontrollers make the design of such battery operated equipment simple and straightforward.

Security patrol systems are another example - the security officer carries a small hand-held data collector which reads codes from the check-points of the patrol route, and a base station collects data at the end of the shift to keep a record of the movements of the officer during the shift. The hand-held unit requires exactly the kind of control that a simple 4 bit microcontroller can provide, and the base station is also a candidate for the use of a more sophisticated model.

Shopfloor time recorders, where workers clock in and out, are becoming increasingly sophisticated as they change from mechanical to electronic operation. They can provide facilities to ease the implementation of flexible working hours, and at the same time provide the management and wages department with the information they require in any easy to use form, perhaps by direct link to their computer. Providing such facilities, and making clocking in and out a quick and efficient operation, is again well within the capabilities of many 4 bit microcontrollers available on the market.

Security is another area where high reliability is often required, both from hardware and software. Proving that a system will not be responsible for false alarms or breaches of security is sometimes difficult. A frequent requirement of such systems will be evidence that the system has been adequately tested in the development phase, and each system may have to be put through a burn-in program to weed out the parts which may fail early in their life.

Industrial applications

Many industrial control systems still rely on hard-wired relay circuits, timers, counters, and mechanical sequence controllers, but electronic control is making inroads into this market as it provides more flexible and reliable operation. The mainstay of such electronic control systems is the programmable logic controller (PLC), of which there are many different types available. The majority of these use 8 or 16 bit microprocessors to provide the required facilities – large numbers of I/O channels, modularity and expandability. However there is no reason why some of the top of the range general purpose 4 bit microcontrollers could not be used to good effect in the smaller PLC units. The main contribution the 4 bit microcontroller has to make in this field, though, is in the peripheral devices to the PLC, such as hand-held programming panels, ROM writers, and data access units. Other applications in the industrial control sector are such things as stand alone temperature controllers, safety systems, counter/timers, liquid flow and level controllers.

Data logging is another important area in industrial control where the 4 bit microcontroller is in its element. Often it is necessary to take readings at regular intervals from a number of sensors and store results, perhaps for onward transmission to a central computing facility when interrogated. When a large number of units are required at remote outstations, perhaps communicating with central control over the telephone system, the tasks involved often come well within the capabilities of a 4 bit microcontroller, and although some special sensing peripherals may be required, the usual range of integrated peripherals makes their use cost effective.

All sorts of areas of control, telemetry, instrumentation and data logging find the application of 4 bit microcontrollers in the industrial field.

Smart sensors

Because of their small size, low power requirements and low cost, 4 bit microcontrollers are ideal for building 'smart sensors'. Such quantities as temperature, pressure, humidity and the sensing of pollutants, require special sensing elements, and special electronic signal conditioning circuits in order to get signals suitable for use with a control system. Often it is desirable to have the signal conditioning electronics physically close to the sensor elements, to cut down problems arising from long lead wires, such as lead resistance (which may vary with temperature) and interference. Having placed some electronics near the sensor, and supplying the power to run it, it opens up the opportunity to enhance the 'front end' by adding a microcontroller. This can then perform A/D conversion, linearisation, perhaps temperature compensation, and a variety of other tasks, depending on the operation, and present the control system with a clean signal over a digital link, immune to all but the most severe electrical interference. It also opens up the possibility of using a fibre optic link for even more immunity from interference. It also gives the opportunity for some local control, with a set point supplied by the central system, and the microcontroller operating a simple algorithm to maintain some particular aspect of the local environment at the required set point.

Metering

Metering is another important area of application for 4 bit microcontrollers, allied to the industrial applications above. Traditional flow meters are purely mechanical, but with the addition of electronics, and particularly with the degree of computing power available with a 4 bit microcontroller, meters can be made more cheaply, and with additional functions. Domestic meters in particular, where a large number are employed, are increasingly using 4 bit microcontrollers in their design. This allows such functions as remote reading to be introduced, and has also made it possible to produce more user friendly prepayment meters. Meters which have card readers built in are much easier to use and much more convenient than the old type requiring coins, and in addition they can be programmed to give a limited amount of credit to ensure that vital supplies are not cut off. Water metering is becoming increasingly commonplace, and this provides an expanding market for new meters, many of which will be designed using 4 bit microcontrollers. With in-built peripherals such as LCD drive, comparator inputs, and real time clock capabilities, the microcontrollers available today make it easy to design in many desirable features, such as tariff rates which change according to the time of day or day of the week. It is even possible to allow tariff rates to be altered by remote control.

Retail trade

There are a number of applications for 4 bit microcontrollers in the retail trade. Electronic cash registers, point of sale terminals, bar code readers, and stock checking equipment can all use these devices. Cash registers can benefit well from the common features found in 4 bit microcontrollers, such as VFD drivers, and the I/O necessary to scan a keyboard and drive a small printer. The printer itself may contain another 4 bit microcontroller handling font generation and motor control. Bar codes are found on nearly every item handled by retailers, and even the smallest outlets can find it convenient to make use of these with a bar code reader as a peripheral to the cash register. Larger stores, where stocking the shelves and keeping track of stock movements from warehouse to point of sale is a complex task, often employ hand-held data units, sometimes with lightpens or similar devices for reading barcodes, to allow these tasks to be carried out quickly and efficiently. Low power 4 bit microcontrollers are ideal for this application.

Medical applications

As with so many other areas of life, medicine has come to rely on electronic equipment to function efficiently and provide the level of service that is expected. Automation has been introduced in a number of aspects, particularly in routine analysis, to handle the volume of testing that has to be carried out. Patient monitoring in one form or another is also carried out automatically by machines, and patients are kept alive by life support systems which rely on electronic control for their vital functions. Such equipment must obviously be highly reliable, and many component manufacturers prohibit the use of their components in life support systems at least without prior notification. A designer is well advised to check on the policy of component suppliers before committing them to the design. Most manufacturers run high reliability screening programmes, and will insist that any parts they supply for critical applications must come through such screening to ensure maximum reliability. One aspect of the use of electronics, and microelectronics in particular, in medicine, is the ability to produce compact, lightweight equipment which allows a patient much more freedom than before. Equipment which used to occupy a trolley beside a hospital bed can now be housed in a pouch strapped to the patient's waist, and perform tasks such as drug administration or monitoring of bodily functions such as blood sugar level, blood pressure, heart rate, etc. Essentially such tasks are similar to industrial data logging and control applications, and just as 4 bit microcontrollers were applicable there, so they are here. Low power consumption and low voltage operation make them ideal for use in miniaturised patient monitoring equipment. Data can be collected continuously at whatever rate is required, and at the end of a predetermined period the data can be unloaded into a personal computer for analysis. Alarm levels can be set to advise the patient that they should take a rest or have some food or take some other remedial or preventive action. Doctors are able to obtain data on the vital functions of their patients under normal conditions, and so get a more realistic picture of their condition.

One other interesting application for 4 bit microcontrollers in the medical field is in the control of artificial limbs. An artificial hand can for example be made with built in 'intelligence', making it much more effective at performing the grasping function.

Aerospace

Another application area where high reliability is essential is the aerospace industry. Aircraft are coming to rely more and more on electronic control, and computers are used for monitoring the status of all sorts of systems within an aircraft, and presenting the information to the flight crew. Fly by wire systems are now employed in the majority of current airliners, and have contributed to safety by reducing crew fatigue, allowing instantaneous checks to be made on pilot actions, preventing the aircraft from being flown into altitudes which would cause excessive stress or instability. The majority of this is carried out by systems with considerably more power than can be supplied by a 4 bit microcontroller, but they do have their part to play.

Weight is always at a premium in aircraft design, and multiplexed wiring systems are used to minimise the weight of cabling wherever possible, and 4 bit microcontrollers can be used at the nodes of such a cabling system to provide the necessary intelligence, as in the automobile system described earlier. 4 bit microcontrollers can also be employed to assist with information display. The amount of information that has to be presented in an aircraft cockpit is vast, and having local intelligence built into individual instruments is a good way of aiding the task of presenting it. The 4 bit microcontroller is an excellent way of supplying local intelligence. With its in-built peripherals, often allied to display driving in one form or another, it provides the functions required compactly, with a minimum of extra peripheral circuitry.

Conclusion

It can be seen from the paragraphs above that 4 bit microcontrollers find their way into almost every aspect of life. Wherever there are electronics there is an opportunity for a 4 bit microcontroller. This chapter is not intended to be an exhaustive examination of the application of 4 bit microcontrollers. Instead it is hoped that it will wet the reader's appetite by providing an overview, a taster, to give some feel for how widely they are employed, and how diverse their capabilities are. 4 bit microcontrollers have been around for a comparatively long time, but manufacturers are still developing their capabilities, taking advantage of advances in processing technology to make them smaller, cheaper, and/or with more and more integrated functions. No doubt the application areas in which they are to be found at work will continue to expand for quite a few years yet.

3 Choosing the right chip - currently available devices

Faced with a possible application for a 4 bit microcontroller, the first task is to select a suitable device from the multiplicity of devices available on the market. The information presented here represents the views of the author and is based on that available from manufacturers. Whilst every effort has been made to ensure that the information is correct neither the author nor the publisher can be held responsible for omissions or misinterpretation.

For a given design there will be a number of imperatives - functions that are essential to the product being designed. This may be a particular display drive, or a certain capability in handling analogue signals, for example. There will also be some desirable qualities, such as low power consumption, on which there will be some flexibility. There will also be some unknown quantities, like how much program space will be required to implement the functions that the product must be able to perform.

On the following pages of this chapter are a number of tables each featuring a different aspect of the specifications of 4 bit microcontrollers, aimed at helping with the process of device selection. Select the table which highlights the primary imperative of the design under consideration, and choose devices which look most likely to be suitable. Then check the other attributes listed in the tables for those devices. This should give a good idea of which manufacturers produce the type of device required, and possibly also indicate the group or family of devices to aim for. It is then necessary to obtain detailed data sheets for the chosen devices from the manufacturers or their agents or distributors. It is only when making a detailed examination of the manufacturer's data sheet that it is possible to make sure that a given device is really appropriate for a particular design. It should be noted that the tables are based on information provided by manufacturers during 1991 and 1992, and that new products and developments are always being introduced. It is wise, therefore, to check a manufacturer's current product list to find out if they have brought out a new part which might suit the application even better.

It is not usually difficult to draw up a short list of parts which will be suitable for a given application, but the final choice may require some detailed investigation of the comparative capabilities of the parts on the short list. Often there will be a critical section of the control program, a time critical routine, or a computationally intense section, which can serve as a bench-mark, and give a feel for how easy it is going to be to program each device for the application. It is often a good idea to try out the instruction sets of competing parts, and write a section of the control program. Sometimes it is only when working at such a detailed level that the shortcomings or advantages of one part over another are discovered.

For a variety of reasons, manufacturers tend to have strengths and weaknesses in their product ranges. It may be that they have traditionally served a particular type of market, and so have a good understanding of the requirements of that market sector, but do not have products which serve other application areas. Sometimes a manufacturer's product range has been developed as a result of demand from other associated companies, and this will tend to cause their range to lean in a particular direction. To give some indication of which manufacturers' products are likely to serve a particular application, there now follows a brief review of each manufacturer's capabilities. Manufacturers are listed in alphabetical order.

Fujitsu

A range of some 44 basic mask programmable models, on the whole well supported with piggyback, EPROM and OTP parts for prototyping and low volume applications. The mask programmable parts are usually available in dual in-line package (DIP) and flat pack versions. Peripherals include LCD and VFD drivers, A/D and D/A converters, and PLL. Memory up to 8192 × 8 bits of ROM and 256 × 4 bits of RAM. Program store is only 8 bits wide, and so some instructions are 2 bytes, or even 3 bytes at the top end of the range. However, the majority of instructions, particularly in the low end of the range, are single byte, single cycle instructions. The instruction sets include subroutines, conditional and unconditional branches, indirect memory addressing, bit manipulation instructions, and in the middle and high range parts, table look up instructions and interrupt handling. Application areas include consumer goods, communications and industrial markets. They have no really low power parts, but some parts are available in a -L (low power) version which will operate from a nominal 3 volt supply and have a typical 30 microwatt standby power consumption.

Hitachi

The H400 series includes 17 different types, all based on the same core processor, and having a variety of different ROM, RAM and integrated peripheral combinations. The mask programmable parts are complemented by a range of EPROM/OTP parts (called ZTAT by Hitachi). For each type there is a choice of high speed or low voltage as well as standard parts. There is also a choice of packaging, including DIL, standard flat pack and a thin flat pack. Integrated peripherals include A/D, D/A (PWM), comparators, PLL, LCD and VFD drive, and DTMF drivers and receivers. ROM sizes range from 2k to 16k by 10 bits, and RAM from 160 to 1184 by 4 bits. The 10 bit ROM word size helps to minimise the number of multi-word instructions, but even so there are some 2-word instructions. The instruction sets include subroutines, conditional and unconditional branches, and table branches from tables held in ROM, direct and indirect memory addressing, bit manipulation instructions, ROM table data look up instructions and interrupt handling. The series is aimed at a wide range of application areas, including consumer goods, telecommunications products, and industrial applications, security systems and automobile use. About half the range are general purpose parts with no special integrated peripherals, but most having plenty of I/O. Some parts are designed for the 'watch' applications but their low power performance does not match up to that of parts by manufacturers who have concentrated on this application area.

Mitsubishi

19 types in two ranges, plus 10 parts designed for use in infra-red remote control transmitters. Piggyback parts are available for most of the general purpose parts, but not for the infra-red devices (though one part in the more general purpose ranges can be used in infra-red transmitters). Packaging choices include DIP, SOP, and QFP. Even one of the piggyback types comes in flat pack, accepting an LCC EPROM. ROM sizes range from 512 × 8 bits to 4k × 9 bits and RAM sizes from 32 × 4 to 256 × 4 bits in the general purpose range. Despite

the limited wordlength in the program store, very few instructions are double-word. These are, however, the branch and subroutine call instructions and so tend to be frequently used. In the general purpose range, all the processors have subroutine handling, indirect RAM addressing, and conditional skips. These devices also include the 'continuous description' technique, whereby an instruction repeated is ignored. For example the LY (Load Y register) instruction can be used repeatedly, each time with a different value. Only the first one encountered has any effect. (NEC use a similar technique, called 'string effect'.) This can be useful for instance in subroutines, providing multiple entry points each with a different parameter setting without the need for jumps or setting the parameter prior to entry. The higher end processors also include interrupt handling, and power down modes. These parts do not include a vast range of integrated peripherals, but do include 8 bit A/D converters, 8 bit serial I/O, LCD drive, PWM, high voltage outputs, and various timers and event counters. Apart from the devices which are specifically designed for use in infra-red transmitters for remote control, the range of parts supplied aims to be of use in a wide range of applications.

National Semiconductor

The COP400 series provides a basic set of 13 parts, together with extended temperature range, low power, high speed, ROMless and piggyback variants. These parts are in NMOS and CMOS technology, and contain 512 to 2k bytes of ROM, 32 to 160 × 4 RAM. The range is not strong on integrated peripherals, but come in small packages, some down to 20 pins. Some of the range will handle interrupts, all include subroutine calls, but nesting is limited by the two level stack in some parts. All parts also include indirect memory addressing, skip on condition, bit manipulation and the 'string effect' on the LBI instruction (load memory pointer with immediate data). The family is aimed at a range of general purpose applications, typically consumer goods, automotive, industrial control, toys, games and telephone applications. A key feature of the National Semiconductor range is claimed to be their low cost. National also offer what they call 'post-metal programming' on some of the parts in their range, a technique which provides customised parts fast, but at lower cost than EPROM or OTP. The National Semiconductor data book also includes some good application notes, which have in them some bright ideas which can easily be transported to other manufacturers' parts.

NEC

Over 60 basic types plus supporting OTP and piggyback versions in several families designed to cover the whole range of applications of 4 bit microcontrollers. Not all parts are supported by equivalent OTP, but most are. Program memory is 8 bits wide and many instructions occupy two bytes. This has the advantage of packing more instructions into the memory at the expense of operating speed. Beware, however, when comparing memory size - it is often specified in words, where each word usually represents one instruction, so a memory size given in bytes may appear bigger, but is in fact smaller. There is a good range of integrated peripherals available, as would be expected from such a large selection of parts, with additions to the range under development. Because the range of parts produced by NEC is so wide, it is more difficult to give meaningful generalisations about facilities available. The parts which drive LCD displays allocate a section of the RAM as display memory. There is no separate display memory as there is in many LCD drive parts. Also, the memory bits are dedicated to specific LCD drive pins which is rather inflexible and can give rise to inefficiencies in program code. The instructions sets in most of the parts include interrupt handling, subroutines, direct and indirect memory addressing, bit manipulation, skip on condition, and the string effect is available on several instructions. To compensate for the large number of 2 byte instructions there are some table look up instructions allowing abbreviated reference to commonly used items, such as subroutines and memory pointers. At the top end of the range, the architectures closely resemble those of 8 bit microprocessors, making it very easy for someone experienced in assembly level programming of the 8 bit parts to transfer to these 4 bit devices. The applications to which the NEC parts can be applied covers the whole range, but they are not strong in extremely low power parts.

OKI

About 20 different basic parts with some OTP and piggyback support, and some ready programmed parts for standard applications such as alarm clocks, calculators and thermometers. RAM capacities go up to 1024 × 4, and ROM to 4k × 15 in one family and 8k × 8 in another. OKI is strong on very low power parts, offering 1.5 volt and 3.0 volt operation, and consuming very few microamps in the Halt state. Integrated peripherals include LCD and VFD drive circuits, melody generators, DTMF generators and A/D conversion. Some of the LCD parts have a good feature in that an area of mask programmable interconnect is used between the display memory and the output pins. This gives complete flexibility to both the software design and the LCD layout, and ensures that the two can be connected by a simple one-to-one pattern on the printed circuit board. Instruction sets in the lower end parts are quite limited, having no subroutine or interrupt handling. They do have Halt Release facilities whereby a change of state on a selected input can cause the processor to come out of a Halt state, which can be used to perform in a similar way to an interrupt system for simple applications. Data tables in ROM are also difficult to use, but the Jump @ instructions can be used to fulfil a similar purpose. With the Jump @ the destination of the jump is determined by the value of a register, typically the accumulator. Bit manipulation is generally easy, and conditional branches are available. RAM addressing can be inflexible in some parts, where pointers can be set to specify a 16 nibble 'page', but within that page all addressing is direct. The principal applications served by OKI are the low power battery operated devices such as watches, calculators, thermometers, etc, and augmented feature telephones. They also have some parts aimed at industrial control applications.

Panasonic

Two series of parts, MN1500 and MN1700. The MN1500 series comprises about 29 devices, all in CMOS, covering a wide range of applications, including telephone, and television applications as well as a number of general purpose parts. There are parts to drive LCD and fluorescent displays, and several with LED drive capability. Analogue signal handling is also catered for in a number of the parts in the series. There are no particularly low power parts, but several devices that operate at reduced voltages. They are well supported by piggyback parts for development use. Memory sizes range from 0.5 kbytes to 8 kbytes for program store, serving a set of up to 124 instructions occupying 1 or 2 bytes each. 65% are single byte instructions. RAM sizes vary from 16 to 512 nibbles, with 6 parts in the range, which are designed for use in telephones, having an extra section of RAM, giving an extra 384 to 1k nibbles. The instruction set contains a useful range of instructions with direct and indirect memory addressing, and auto incrementing or decrementing of the memory pointers; a stack in RAM for interrupt and subroutine handling, and for general purpose use; decimal adjust; single bit manipulation instructions, including single bit output operations, and a pulsed output, to give short pulses (quarter machine cycle, typically 500 nanoseconds). There are also a number of instructions allowing 8 bit data manipulation, including load, store, shift, increment, decrement, input and output. It is also possible to access table data in ROM. A good selection of jumps are included in the instruction set, including, unusually, a 'Compare and Jump' instruction. Note that the lower end devices do not have the complete instruction set available, but the data book does not appear to indicate which instructions are not available.

The MN1700 series is a group of 7 parts designed with higher performance in mind. These are also supplemented by piggyback versions for user programming. The applications for this series will overlap with those for 8 bit parts, and there are a number of 8 bit features built into the MN1700 devices, including easy 8 bit transfers to and from memory, and to and from I/O ports. The series has been designed to allow the use of external memory as an extension of internal memory, and the interrupt system allows up to 11 sources of interrupt in some members of the family.

Sanyo

Fifteen families of devices with a further five families under development, each family including a range of parts with varying ROM and RAM sizes. There is a good selection of parts for driving LCDs, 63 to 140 segments, all available in chip form as well as packaged. They are not however supported by user programmable parts, there being only the LC58E68, a windowed EPROM part. The remaining parts are well supported by piggyback, OTP and EPROM versions. ROM sizes vary from 512 bytes to 16k bytes, either as 8k × 16 or 16k × 8. The parts which have 16 bit wide program memory all have single word instructions, but the parts with 8 bit wide program memory have single and double byte instructions, and also the occasional 3 byte instruction. RAM capacities go from 32 × 4 to 608 × 4. Instruction sets vary across the range of parts. At the low end are simple parts with no interrupt handling, but with halt release, and single level subroutine nesting. At the top end there is multiple interrupt handling, many level subroutine nesting, push and pop instructions, and table generated jumps. Table look up instructions are common to most parts, as are conditional branches and indirect memory addressing. The architecture of some of the LCD drive parts is designed to make them particularly useful in watches and other applications requiring accurate time-keeping, with the appropriate timebase dividers, melody generators and good low power modes. One part is designed for use in remote control transmitters, with some further parts under development. The non-LCD parts have a variety of peripherals on board, including high voltage drive capability, A/D inputs and D/A converters.

Seiko Epson

A single family with 13 different controllers based on the same core processor. There are variants of several parts for low voltage operation and higher speed. This is a low power family, most of the parts operating from a nominal 3.0 volts in the standard form, 1.5 volts in the low voltage variant. At the lowest power end of the range there is a part which will operate at a typical power of 8 microwatts. Memory sizes are from 1024 to 6144 × 12 bits for the ROM with no multi-word instructions, and 80 to 1024 × 4 bits for the RAM. There are no user programmable parts in the family. Integrated peripherals include LCD drivers on all but two parts in the range, with one part designed for a dot matrix LCD, capable of driving 640 elements (40 × 16). Also to be found in the range are A/D converters, remote control carrier generation, comparators, event counters, sound generators, external memory support, serial I/O-UART (universal asynchronous receive/transmit), and low supply voltage detection. Parts are packaged in QFP or are available as naked chips. Several members of the family are dual clock parts, so instruction times change according to the mode that is in use. At the slow rate, execution is 153, 214, or 366 microseconds, depending on the instruction in use, and at the fast rate the equivalent figures are 10, 14 or 24 microseconds typically, with some parts operating as fast as four times this speed. The instruction set contains a basic 100 instructions, with a further 8 available in some parts. The set supports subroutine calls (using a stack), and interrupt handling (mostly 2 or 3 external with several internal interrupts depending on chip function). Bit manipulation, unusually for a 4 bit microcontroller, is not catered for explicitly, but has to be carried out using AND, OR and XOR instructions. There is no string effect available. Jumps, conditional and unconditional, are available, together with a programmable jump. Memory addressing is both direct and indirect, but it does not appear to be possible to address tables in ROM. I/O is mapped into the RAM address space which has a 12 bit address. The family of parts is aimed primarily at the very low power, battery operated device market - watches and timers of all sorts, games, and electronic diaries, dictionaries, and pocket translators being possible with the part which has sufficient I/O to allow access to 1Mbit of external memory. Two parts in the family are aimed at infra-red remote control applications, and two parts are aimed at the remote metering task, featuring extended serial I/O facilities.

SGS-Thomson

A second source for the National Semiconductor COPS range of microcontrollers, with the addition of some extra packaging options, namely SOIC and PLCC. The SGS-Thomson range does not offer a second source for every part in the National Semiconductor range, but the parts that are covered are almost identically specified. There appear to be minor differences, for instance in output current capabilities, and if a second source is a system requirement, the user must examine the specifications carefully to ensure total suitability of the second source.

Sharp

A range of 21 parts aimed at a variety of applications, all but one being mask programmable. These parts are aimed at volume markets and do not in general have the OTP equivalents for low volume runs. ROM sizes range from 508 × 8 to 8192 × 9, with a high percentage of the instructions taking only one word of ROM. RAM sizes range from 32 × 4 up to 320 × 4. Nine members of the Sharp range support LCD drive, and most of these have a separate section of RAM to hold display data, with capabilities extending up to 200 elements of display. There is a fixed relationship between RAM location and LCD segment pin, an arrangement which is fairly common but inflexible, and can cause problems in the software, the LCD layout, or the printed circuit board layout which provides the interconnection between the two. The software designer has to be aware that decisions made to improve efficiency of code may make life difficult for the LCD display designer. The 1.5 volt parts have internal circuitry for generating 3.0 volts to drive the LCD in a ½ duty, ½ bias mode, but some nominal 3.0 volt parts require 2 × 1.5 volt cells to generate the necessary LCD voltages. If such parts were to be powered from, say, a lithium cell then external resistors would be required to generate the centre voltage for the LCD. Where the LCDs to be driven are ¼ duty, ⅓ bias the necessary voltage divider resistors are included on chip.

Other peripherals supported in the range include a VFD driver, serial I/O, A/D conversion and some high current outputs. Some of the LCD drive parts also include melody generators (together with a section of dedicated ROM). Devices come packaged in DIPs, skinny DIPs, and quad flat packs. Software capabilities vary across the range, the number of different instructions available varying from 41 to 98. All parts support subroutine calls, though some of the simpler LCD drive parts only allow a single level of subroutine nesting. The more capable parts have 4 or 6 level nesting, and some parts use the RAM for a stack. Some parts support interrupt handling, typically from 2 external and 3 internal sources. The parts which do not have interrupt handling have a standby release facility with certain inputs. Note however that the standby mode usually involves stopping the oscillator, and there may be a delay before processing can begin again after standby is released. All parts in the range include bit manipulation instructions, conditional skips, and jump instructions. The string effect applies to the LAX (load accumulator) instruction for many parts in the range. Some of the more powerful members of the range include table look up and programmable jump instructions. All the devices have indirect memory addressing, some at the top end also have direct addressing. The LCD drive parts are mainly low power parts aimed at battery operated devices such as watches, calculators and games. Two of the range operate from 1.5 volts and consume typically 10 to 12 microamps when operating, making them ideally suited for this type of application. Most of the LCD drive parts operate from 32 kHz crystal oscillators, and have cycle times of 61 microseconds. The non LCD parts are all CMOS parts operating from 3 or 5 volt supplies, consuming typically 1 to 15 milliamps over the range of parts. They are designed for general purpose control applications such as domestic environmental controls and other situations where compact low power intelligence is required. Cycle times are in the 1 to 2.5 microsecond range, offering higher performance than many manufacturers parts, with oscillators controlled by resistor or ceramic resonator. Since this is not suitable for accurate time keeping, some parts have provision for the addition of a 32 kHz crystal as well.

Texas Instruments

The TMS 1000 family is now obsolete, and not being used for new designs. As a result data from the family has not been included in the tables. The TSS400 family is a range of sensor signal processors aimed at temperature, pressure and acceleration measurement. They include LCD drive, and operate on low power, making them suitable for battery operated instruments. There is no EPROM version, but a preprogrammed version, containing an interpreter, allows the user to store control code in an external serial EEPROM. Suitable for low volume applications, but not suitable for checking the code for other members of the TSS 400 family prior to committing to mask charges. Texas Instruments also produce a family of 4 bit microcontrollers designed for telephone applications, and termed 'telephone microcomputers'. These parts, the TCM8305B to 8B, operate from a wide range of voltages, 2.5 to 5.5 volts at fairly low power, working with an instruction set of 80 instructions. They have provision for battery back-up to the RAM, maintaining memory contents with battery voltage as low as 2 volts, supplying 10nA typical standby current. Texas Instruments seem very shy about these parts, as neither the TSS 400 nor the TCMB 830xB family appears in the 1990 Master Selection Guide (the most recent one available at the time of writing).

Toshiba

A comprehensive range of 80 parts in 5 families (some still under development) together with OTP and piggyback equivalents for most parts. At the bottom end of the range is a family of low cost parts in small packages with 512 × 8 and 1k × 8 ROM and 32 nibbles of RAM. The remaining families cover the range of ROM sizes from 1k to 16k × 8 and 128 to 1024 nibbles of RAM. The instruction set for the more powerful families includes quite a few 2-byte and some 3-byte instructions, but the instructions for the simplest devices are all single byte. As is to be expected from a range of parts like this, the selection of integrated peripherals is quite wide. For display drives, there is the capability of driving VFT, LED, on-screen TV display up to 80 characters (chosen from a character set of up to 128 characters) and LCD, with one part capable of driving a 440 dot display. The LCD drivers feature on-chip bias generation and an external input which allows the use of a display whose operating voltage is different from the microcontroller operating voltage. Other features include zero crossing detection, various types of A/D converter, several interval/event timers, remote control preprocessors, comparators, a UART, key-scan functions, DTMF receivers and transmitters, PWM output, 'Beep' output, and a pulse generator. Parts are packaged mainly in skinny DIP or quad flat pack, with a few of the smaller parts in standard

DIP, and a few in small outline package. Instruction times vary from 1 to 2.5 microseconds, and some parts are dual clock, allowing a 32 kHz watch crystal to be used for accurate timing functions.

The main families claim to be software compatible, sharing an instruction set with a basic 90 instructions, extending to 105 instructions for some members. Facilities available include subroutine calls, using a stack for return address, interrupt handling (most parts handle 2 external and 4 internal interrupts), bit manipulation instructions, conditional branches and table look up instructions. Both direct and indirect RAM addressing is available. There is also an extended instruction set giving access to extra conditional branches and subroutine calls, and some shift instructions. These expand on assembly to typically 3 or 4 standard instructions. In addition to these facilities there is a compiler which provides a number of higher language facilities such as IF and WHILE, and allows programs to be structured in an orderly fashion. Applications for this range of devices cover the whole range of jobs for which 4 bit microcontrollers are suited, except perhaps the very lowest power devices, but even here performance is only bettered by the companies who have specialised in the low power end of the market.

Selection tables

A word of warning – The figures quoted in the tables are taken from manufacturers' data books and data sheets. In general typical figures have been quoted as they usually give the best idea of performance over a large population of devices, but cannot be relied on. One problem is that manufacturers make different assumptions about the operating conditions for their typical values, so that the values given by different manufacturers are not truly comparable. However, looking at maximum or minimum figures is not always reliable for comparison purposes either. Some manufacturers' maximum figures seem to be much further from typicals than others, presumably to achieve acceptable yield levels from their process. For example, if you are designing a piece of battery powered equipment and you want to quote battery life, to use the maximum current consumption figure would give a very pessimistic value. Only under extreme temperature conditions will a small proportion of the parts ever reach the maximum value quoted. It may be necessary to take account of this in the design, but to assume that this will be the normal steady state condition is unrealistic. The typical current figure will give a much better estimate of performance under normal conditions, and if the maximum were to be a serious embarrassment, the manufacturer will probably be persuaded, for a premium, to screen out parts which exceed a chosen value.

The tables concentrate on the following aspects:

Table 1 Low operating voltage - devices which operate at nominal 1.5 and 3.0 volts. If devices are not listed here they should be assumed to operate at 5.0 volts though there are a few higher voltage parts available. Note that manufacturers sometimes specify operating supply voltages over a range, say 3 to 5 volts, but also indicate that some chip functions, for example analogue I/O, will only operate at 5 volts. This is not highlighted in the table, so watch out for it in the data books.

Table 2 Low power consumption - devices which operate at the lower end of the power range, less than 100 microwatts (typical values unless indicated otherwise). Note that the majority of 4 bit microcontrollers are fabricated in CMOS technology, and so are inherently low power, especially at low speed. This table just selects parts which are specified at the extreme end of the power spectrum.

Table 3 Display drive - devices which have integral drivers for LCD, VFD, or TV on-screen displays. Some manufacturers list LED drive capability in their data sheets, but the current capacity required is common, so LED drive has not been included here.

Table 4 Analogue I/O - devices which have in-built comparators, A/D converters and PWM O/P.

Table 5 Serial I/O - devices which include provision for serial data transfer. The maximum clock rate given is usually with an internal clock, and is relative to the main CPU oscillator clock, i.e. the frequency as seen at the oscillator terminals. All serial I/O is synchronous unless otherwise indicated.

Table 6 Parallel I/O - every part available gives some provision for parallel I/O. This table highlights parts which allow more than 4 bits to be output in a single instruction, or parts which provide special facilities such as strobed I/O.

Table 7 DTMF drivers and receivers - devices which are designed primarily for the telecommunications market, but can find application in other fields as well.

Table 8 Phase locked loops – devices which include PLL for auto tuning systems.

Table 9 Memory provision – an indication of the ROM and RAM provided, also parts with EEPROM non-volatile memory on board.

Table 10 Computing power – an indication of the power of the various manufacturers' families of microcontrollers, giving the typical instruction cycle time. Note that this is usually a 'best' figure, and that the power available depends on the operating conditions, especially the operating voltage. If a part can operate at say 5 volts or 3 volts, then the figure in the table will refer to 5 volt operation. The performance at 3 volts will be considerably slower.

Table 11 Special facilities – a table designed to highlight some of the other facilities available such as melody generators and watch-dog timer provision.

Table 12 Packaging – the packaging options, including devices available as bare chips for use where space is at a premium, EPROM versions for development and OTP for low volume.

Table 13 Second sourced parts.

Note: Manufacturers tend to produce parts in families, often the same processor with different amounts of ROM, and the numbering of these parts is related. Where a whole family of parts is referred to in the table a lower case 'x' is used as a 'wild card' in the part number, e.g. the devices MSM5052, MSM5054, MSM5055 and MSM5056 will be referred to collectively as MSM505x. Packaging information is also often included in the part number by a code letter or letters on the end. The wild card used in this case is an upper case 'X'. Sometimes these have been omitted for the sake of clarity.

In some cases information supplied by manufacturers has not been as complete as is necessary to fill in all the spaces in the tables. Where a facility is provided, but details of the provision are not known, ?? appears in the table.

Table 1 Low voltage parts

Fujitsu	
MB885xLM	2.5–4.0
Hitachi	
HMCS40xCL	2.7–6.0
HMCS41xC	3.5–6.0
HMCS41xCL	2.5–6.0
HMCS424C	3.5–6.0
HMCS424CL	2.5–6.0
HD40L4222	2.5–6.0
HD404328	2.7–6.0
HD404608	2.7–6.0
HD404618	2.7–6.0
HD404720	2.7–6.0
HD40L4808	2.7–6.0
Mitsubishi	
M34200M4	2.5–5.5
M34201M4	2.2–3.6
M34210M2	1.1–3.6
M5046x	2.2–5.5
M5056x	2.2–3.6
National Semiconductor	
COP3xxC	3.0–5.5
COP410C	2.4–5.5
COP411C	2.4–5.5
COP413C	3.0–5.5
COP42xC	2.4–5.5
COP44xC	2.4–5.5

NEC

All NEC 4 bit microcontrollers are specified for operation over the range 2.7 to 5.5 volts or wider, except the 7533 which is specified from 3.0 to 6.0 volts. Performance is reduced at lower voltages.

OKI	
MSM5052	1.25–1.65
MSM5054	1.25–1.65 or 3.0 V operation mask selectable
MSM5055	1.25–1.65 or 3.0 V operation mask selectable
MSM5056	1.25–1.65 (solar cell)
MSM6051	1.25–1.65 or 3.0 V operation mask selectable
MSM6351	1.25–1.75 or 2.6–3.5 (mask option)
MSM6353	1.25–1.75 or 2.6–3.5 (mask option)
MSM6352	2.0–5.5
MSM6512	2.4–3.6
MSM6545	0.9 V min
MSM6546	0.9 V min
MSM6575	0.9 V min
Panasonic	
MN1550	2.5–5.5
MN15514	2.5–5.5
MN1551A	2.5–5.5
MN15522	2.5–5.5
MN1554	3.0–5.5
MN15542	2.5–5.5
MN150401	2.2–5.5
MN150402	2.2–5.5

Table 1 cont.

MN158281	2.5–5.5
MN158486	2.5–5.5
MN1584532	2.0–5.5
MN158416	2.0–5.5
MN158614	2.5–5.5
MN158655/855	2.0–5.5
MN158851	2.5–5.5
MN15834A	2.5–5.5
MN158341	2.5–5.5
MN158412	2.5–5.5
MN158413A	2.5–5.5
MN158417	2.0–5.5
MN158810	2.0–3.6
MN17541	3.0–5.5
MN17581	3.0–5.5
MN171601	3.0–5.5
MN178122	2.7–5.5
MN178611	2.7–5.5
Sanyo	
LC5732	1.3–1.65 or 2.0–3.6 (mask option)
LC5733	1.3–1.65 or 2.6–3.6 (mask option)
LC5734	2.0–6.0
LC5851	1.3–1.65 or 2.6 to 3.6 (mask option)
LC586xH	2.0–6.0 (2.0 V not available for all LCD drive options)
LC6520C	3.0–5.5
LC6522C	3.0–5.5
LC6527C	3.0–6.0
LC6528C	3.0–6.0
LC6538D	2.7–6.0 (Low voltage at reduced operating speed)
LC6543C	3.0–6.0
LC6546C	3.0–6.0
LC6510xA	3.0–6.0
LC65204A	2.7–6.0
Seiko Epson	
SMC621x	2.2–3.5
SMC62L3x	0.9–1.7
SMC623x	1.8–3.5
SMC62A3x	2.2–3.5
SMC624x	2.2–5.5
SMC6251	0.9–1.8
SMC62L51	1.8–3.6
SMC6262	2.2–5.0
SMC62L62	0.8–2.2
SMC62A62	2.2–5.0
SMC6266	2.2–3.5
SMC6274	2.4–5.5
SMC628x	1.8–3.5
SMC62L8x	0.9–3.5
SMC62A8x	1.8–3.5
SMC62B81	0.9–3.5
SGS-Thomson	
ETC93xx	3.0–5.3
ETC9410	2.4–5.5
ETC9411	2.4–5.5

Table 1 cont.

ETC942x	2.4–5.5
ETC944x	2.4–5.5
Sharp	
SM4A	2.6–3.2
SM500	2.7–3.3
SM51x	2.6–3.2
SM53x	1.2–1.8
SM55x	2.7–5.5
SM563	2.7–5.5
SM57x	2.7–5.5
SM59x	2.5–5.5
SM5E4	2.7–5.5
SM5J6	2.7–5.5
SM5K1	2.4–5.5
Texas Instruments	
TCM8305B/6B	2.5–5.5
TCM8307B/8B	2.5–5.5
TSS400	2.6–5.5
TSS403	2.6–3.8
TSS400/4	2.6–5.5
TSS403/4	2.6–3.8
Toshiba	
TMP47C855	2.2–6.0
TMP47C858	2.7–6.0
TMP47C451A	2.2–6.0
TMP47C452A	2.2–6.0
TMP47C453A	2.2–6.0
TMP47C454A	2.7–6.0
TMP47C456A	2.7–6.0
TMP47C25	2.5–6.0
TMP47C26	2.2–4.0

Table 2 Low power devices

Fujitsu
No parts in the <100 microwatt range.

Hitachi		
HD404608/18	35 μA @ 3 V	– 'Sub-active' Slow speed limited operation.
	15 μA @ 3 V	– 'Watch mode' Stopped, waiting for timer, LCD active.
HD40L4808	as 4608	
HD404439	as 4608	
HD404719	as 4608	
HD404728	as 4608	
HD404729	as 4608	
Mitsubishi		
M34210M2	80 μA @ 1.1 V	

National Semiconductor
No parts operating at <100 microwatts.

Table 2 cont.

NEC		
μPD7506	35 μA @ 3 V	
μPD7500x	30 μA @ 3 V	(low speed mode)
μPD753xx	30 μA @ 3 V	(low speed mode)
OKI		
MSM5052	3 μA @ 1.5 V	(temperature sampling off)
MSM5054	3 μA @ 1.5 V	
MSM5055	3 μA @ 1.5 V	
MSM5056	3 μA @ 1.5 V	
MSM6051	3 μA @ 1.5 V	
MSM6351	3 μA @ 3.0 V	
MSM6353	3 μA @ 1.5 V	
MSM6512	30 μA @ 3 V	

Panasonic
No parts operating at <100 microwatts.

Sanyo
It is difficult to tell from the data book the exact conditions under which some of the quoted figures apply.

LC5732	3 μA
LC5733	3 μA
LC5734	3 μA
LC582x	5 μA
LC5851	3 μA/10 μA
LC586xH	20 μA
Seiko Epson	
SMC621x	3.0 μA (HALT)
SMC6232	1.2 μA (HALT)
SMC62L32	1.0 μA (HALT)
SMC62A32	3.0 μA (HALT)
SMC6233	1.2 μA (HALT)
SMC62L33	1.0 μA (HALT)
SMC62A33	1.5 μA (HALT)
SMC6235	1.8 μA (HALT)
SMC62L35	1.5 μA (HALT)
SMC62A35	2.0 μA (HALT)
SMC6237	2.5 μA
SMC623A	2.5 μA
SMC6244	3.0 μA (HALT)
SMC6246	2.5 μA (HALT)
SMC6248	3.0 μA (HALT)
SMC6251	3.5 μA
SMC6262	7.0 μA
SMC62L62	5.0 μA
SMC6266	1.8 μA (HALT)
SMC6274	7.0 μA
SMC6281	2.5 μA
SMC62L81	2.5 μA
SMC62A81	7.2 μA
SMC62B81	7.2 μA
SMC6282	4.5 μA

SGS-Thomson
No parts operating at <100 microwatts.

Table 2 cont.

Sharp		
SM530	12 μA @ 1.5 V	
SM531	10 μA @ 1.5 V	
SM500	20 μA @ 3 V	
Texas Instruments		
No parts operating at <100 microwatts.		
Toshiba		
TMP47C855	30 μA @ 3 V	(slow mode)
TMP47C858	30 μA @ 3 V	(slow mode)
TMP47C456A	15 μA @ 3 V	(slow mode)

Table 3 Display drive capabilities

	VFD segments	LCD segments	TV
Fujitsu			
MB885xB	8×8	–	–
MB88205B	8×8	–	–
MB88514B	7×18	–	–
MB88515B	7×18	–	–
MB88516B	7×10	–	–
MB88517B	6×8	–	–
MB88525B	8×16	–	–
MB88526B	8×16	–	–
MB88543	–	32×4	–
MB88544	–	24×4	–
MB88545	–	32×4	–
MB88546	–	24×4	–
MB88561	–	26×2	–
MB88562	39	–	–
MB88563	–	26×2	–
Hitachi			
HD40430x	25	–	–
HD40432x	–	24×4	–
HD4046x8	–	32×4	–
HD404708	16×16	–	–
HD404709	16×16	–	–
HD404720	16×16	–	–
HD404728	16×16	–	–
HD404729	16×16	–	–
HD404808	–	32×4	–
Mitsubishi			
M50723	–	23×4	–
M50922	–	27×4	–
M34200M4	–	25×4	–
M34201M4	–	35×4	–
M50565	–	32×3	–
M50433B	–	–	16 chars
M50435	–	–	16 chars

Table 3 cont.

	VFD segments	LCD segments	TV
M50436	–	–	48 chars
M50439	–	–	64 chars
M50442	–	–	32 chars
M34300M4	–	–	32 chars
M34300N4	–	–	32 chars
M34350N6	–	–	32 chars
National Semiconductor			
No integrated display drivers.			
NEC			
μPD7501	–	24×4	–
μPD7502	–	24×4	–
μPD7503	–	24×4	–
μPD7514	–	32×4	–
μPD7516H	8×16	–	–
μPD7519(H)	8×16	–	–
μPD75206/8	10×12	–	–
μPD7521A/6A	14×12	–	–
μPD75236/7/8	24×9	–	–
μPD75268	14×12	–	–
μPD753xx	–	32×4	–
μPD75328	–	20×4	–
μPD1708	–	??	–
μPD171x	–	??	–
μPD17001/6	–	30×2	–
μPD17051	–	–	97 chars
μPD17053	–	–	199 chars
μPD17102	–	12×4	–
μPD172xx	–	34×4	–
OKI			
MSC6458	12×12	–	–
MSM5052	–	26×2	–
MSM5054	–	44×2	–
MSM5055	–	60×2	–
MSM5056	–	38×2	–
MSM6051	–	63×3	–
MSM6351	–	58×4	–
MSM6442	–	46×2	–
MSM6512	–	27×4	–
MSM64162/4	–	??	–
Panasonic			
MN158682/882	24	–	–
MN158418	41	–	–
MN1584532	–	32×4	–
MN158416	–	8×3	–
MN158614	–	32×4	–
MN158655/855	–	32×4	–
MN158851	–	29×4	–
MN15151	–	–	2×16 chars
MN15287	–	–	6×16 chars
MN152810	–	–	6×16 chars
MN152121	–	–	6×20 chars

Table 3 cont.

	VFD segments	LCD segments	TV
MN178611	-	32×4	-
MN178122	16×16	-	-
Sanyo			
LC573x	-	27×3	-
LC582x	-	42×4	-
LC5851N	-	25×3	-
LC586x	-	35×4	-
LC6512A/13A	10×8	-	-
LC6514B/15B	10×8	-	-
LC6538D	12×12	-	-
LC6554D/H	13×8	-	-
LC6568D/H	13×8	-	-
LC67216A	16×12	-	-
Seiko Epson			
SMC621A	-	32×4	-
SMC6215	-	50×4	-
SMC6232/L32/A32	-	38×4	-
SMC6233/L33/A33	-	40×4	-
SMC6235/L35/A35	-	48×4	-
SMC6237/L37	-	26×4	-
SMC6244/46	-	40×16	-
SMC6248	-	51×16	-
SMC6251/L51	-	26×4	-
SMC6274	-	32×4	-
SMC62x81	-	26×4	-
SMC62x82	-	42×4/38×8	-
SMC4202	-	16×3	-
SGS-Thomson			
No integrated display controllers.			
Sharp			
SM5J5	16×10	-	-
SM530	-	48×2	-
SM531	-	40×2	-
SM500	-	28×2	-
SM5K1	-	16×4	-
SM4A	-	34×2	-
SM510	-	33×4	-
SM511	-	34×4	-
SM512	-	50×4	-
SM563	-	32×4	-
Texas Instruments			
TSS400 Family	-	20×4	-
TSS400-S230		7 digits	
Toshiba			
TMP47C111N/M	16 bits	-	-
TMP47C211N/M	16 bits	-	-
TMP47C1270N	12×16	-	-
TMP47C1670N	12×16	-	-
TMP47C1237N	-	-	20×4 chars

Table 3 cont.

	VFD segments	LCD segments	TV
TMP47C1637N	–	–	20×4 chars
TMP47C1238N	–	–	20×4 chars
TMP47C1638N	–	–	20×4 chars
TMP47C620F	–	32×4	–
TMP47C820F	–	32×4	–
TMP47CE820F	–	32×4	–
TMP47C824F	–	21×4	–
TMP47C434N	–	–	16×2 chars
TMP47C634N/F	–	–	16×2 chars
TMP47C834N	–	–	16×2 chars
TMP47C635N	–	–	16×2 chars
TMP47C647F	–	32×4	–
TMP47C847F	–	32×4	–
TMP47C855F	–	32×4	–
TMP47C858F	–	44×10	–
TMP47C662N	27 bits	–	–
TMP47C862N	27 bits	–	–
TMP47C670N	12×16	–	–
TMP47C870N	12×16	–	–
TMP47C210AN/F	8×12	–	–
TMP47C410AN/F	8×12	–	–
TMP47C212AN	8×12	–	–
TMP47C412AN	8×12	–	–
TMP47C220AF	–	24×4	–
TMP47C420F	–	24×4	–
TMP47C221AF	–	24×4	–
TMP47C421AF	–	24×4	–
TMP47C423AF	–	24×4	–
TMP47C425AF	–	24×4	–
TMP47C236AN	–	–	3 chars
TMP47C336AN	–	–	3 chars
TMP47C237AN	–	–	bar
TMP47C337AN	–	–	bar
TMP47C441AN/F	8×8	–	–
TMP47C446AF	–	24×4	–
TMP47C456AF	–	32×4	–
TMP47C475AN	16×8	–	–

Table 4 Analogue I/O capabilities

	A/D bits × channels	D/A (PWM) bits × channels
Fujitsu		
MB88211	8×1	
MB88511	8×4	
MB88512	8×8	
MB88513	8×4	
MB88514B	8×8	
MB88515B	8×8	
MB88516B	8×8	

Table 4 cont.

	A/D bits × channels	D/A (PWM) bits × channels
MB88517B	8×4	
MB88535		6×3
MB88536		6×3, 13×1
MB8855x(H)	5×4	
MB8856x	6×3	
MB88572	8×8	
Hitachi		
HD404222	I/P comparator × 2	
HD40430x	8×4	
HD404328/9	8×4	
HD404408/18	I/P comparator × 4	8×3
HD404439	8×8	
HD404608/18	I/P comparator × 2	
HD404719	8×8	
HD404728/9		14×1
HD404720		14×1
HD404808	I/P comparator ×2	
Mitsubishi		
M50723	8×4	
M50725/M50925	8×2	
M50726	8×4	
M50727	8×4	
M50927	8×2	
M34200M4	8×8	
M34201M4	8×4	
M34203M4		8×1
M50430		14×1
M50431		14×1, 6×3
M50432		14×1, 6×4
M50433B		14×1, 6×1
M50434		14×1
M50435		14×1, 6×1
M50436		14×1, 6×3
M50439		14×1, 6×8
M50440	3×1	
M50442	3×1	6×4
M50443	3×1	6×4
M50445	3×1	6×1
M34300M4	3×1	14×1, 6×1
M34300N4	3×1	14×1, 6×1
M34350N6	3×1	6×9
National Semiconductor		
No integrated analogue signal handling.		
NEC		
μPD7533	8×4	
μPD75512/6	8×8	
μPD75028	8×8	
μPD75236/7/8	8×8	
μPD75206/8		14×1
μPD75212/16		14×1
μPD75328	8×6	

Table 4 cont.

	A/D bits × channels	D/A (PWM) bits × channels
μPD1700x	4 × 6	6 × 4
μPD17051/3	4×8	14×1, 6×3
μPD17102		6×1
OKI		
MSM5052	Thermometer	
MSM6434	8×4	
Panasonic		
MN158281	8×4	8×2
MN1258486	8×4	8×2
MN158682/882	8×8	10×1
MN15151	5×4	6×5
		7×4, 14×1
MN15287	4×1	6×10, 14×1
MN152810	5×4	6×5
		7×4, 14×1
MN158381		8×2
Sanyo		
LC6530/2		6×3, 14×1
LC65102A/4A	8×8	
LC65204A	8×8	
LC6568D/H	I/P comparator × 4	
LC6630xA	I/P comparator × 4	
LC6640xA	6×6	
LC665xxB	I/P comparator × 4	
Seiko Epson		
SMC621A	I/P comparator × 1	
SMC6215	I/P comparator × 2 (mux)	
SMC6232/L32/A32	I/P comparator × 1	
SMC6233/L33/A33	I/P comparator × 1	
SMC6235	I/P comparator × 1	
SMC6237	I/P comparator × 1	
SMC6251	Temperature by resistance to frequency conversion	
SMC6266	I/P comparator × 2	
SMC6274	5 channel dual slope A/D	
SMC6281/L81	I/P comparator × 1	
SMC4202	6×7 Succ. Approx/Dual Slope (3200 count)	
SGS-Thomson		
No integrated analogue signal handling.		
Sharp		
SM578/9	8×20	
SM5J5/6	8×10	
Texas Instruments		
TSS400 family	12×4	
Toshiba		
TMP47C241	8×4	
TMP47C1260/1660	8×8	
TMP47C1237/1637	I/P comparator × 4	7×7, 14×1
TMP47C1238/1638	I/P comparator × 4	7×9, 14×1

Table 4 cont.

	A/D bits × channels	D/A (PWM) bits × channels
TMP47C434/634	I/P comparator × 3	6×4, 14×1
TMP47C834/635		6×4, 14×1
TMP47C647/847	8×8	
TMP47C850	I/P comparator × 4	
TMP47C660/860	8×8	
TMP47C662.862	8×8	
TMP47C670/870	I/P comparator × 4	14×1
TMP47C231A	4×5	6×1
TMP47C232A		6×3, 14×1
TMP47C233A/433A	3×1	6×3, 14×1
TMP47C236A/336A	3×1	6×1
TMP47C237A/337A	3×1	6×1, 14×1
TMP47C440	8×8	
TMP47C441/242	8×4	
TMP47C446	8×8	
TMP47C475		14×1

Table 5 Serial I/O capabilities

	Bits	Clock	Max rate
Fujitsu			
MB885x,/B,/F,/H	4	INT/EXT	Clock/6
MB885xx	4/8	INT/EXT	Clock/24
Hitachi			
HMCS402	8	INT/EXT	Clock/8
HMCS404	8	INT/EXT	Clock/8
HMCS408	8	INT/EXT	Clock/4
HD404019	8	INT/EXT	Clock/4
HMCS424	8	INT/EXT	Clock/4
HD404222	8	INT/EXT	Clock/4
HD404328/9	8	INT/EXT	Clock/8
HD404408/418	2×8	INT/EXT	Clock/8
HD404439	2×8	INT/EXT	Clock/4
HD404608/18	8	INT/EXT	Clock/4
HD404678	2×8	INT/EXT	Clock/8
HD404708/9	8	INT/EXT	Clock/4
HD404719	2×8	INT/EXT	Clock/4
HD404720	8, 8/16	INT/EXT	Clock/4
HD404728/9	8, 8/16	INT/EXT	Clock/4
HD404808	8	INT/EXT	Clock/4
Mitsubishi			
M50721	8	INT/EXT	Clock/8
M50722	8	INT/EXT	Clock/8
M50725/925	8	INT/EXT	Clock/8
M50726	8	INT/EXT	Clock/8
M50727	8	INT/EXT	Clock/8
M50927	8	INT/EXT	Clock/8
M34200M4	8	INT/EXT	Clock/8

Table 5 cont.

	Bits	Clock	Max rate
National Semiconductor			
COP2xx	4	INT	Clock/4
COP3xx	4	INT	Clock/4
COP4xx	4	INT	Clock/4
NEC			
μPD7501	8	INT/EXT	Clock/2
μPD7502/3	8	INT/EXT	Clock/2
μPD7507/8	8	INT/EXT	Clock/2
μPD7514	8	INT/EXT	Clock/2
μPD7516/9	8	INT/EXT	Clock/16
μPD7527/37/28/38	8	INT/EXT	Clock/2
μPD7533	8	INT/EXT	Clock/2
μPD7554/64	8	INT/EXT	Clock/2
μPD751xx	8	INT/EXT	Clock/4
μPD752xx	8	INT/EXT	Clock/4
μPD755xx	2×8	INT/EXT	Clock/8
μPD7500x	8	INT/EXT	Clock/8
μPD753xx	8	INT/EXT	Clock/8
μPD75402	8	INT/EXT	Clock/16
μPD17102	8	INT/EXT	Clock/16
OKI			
MSM6351/3	5/8 (async)	INT/EXT	Clock/1
MSM6404	8	INT/EXT	Clock/4
MSM6408	8	INT/EXT	Clock/4
MSM6411	8	INT/EXT	Clock/4
MSM6458	8	INT/EXT	Clock/4
Panasonic			
MN1551A	8	INT/EXT	
MN15522	8	INT/EXT	
MN158381	8	INT/EXT	
MN1554	8	INT/EXT	
MN15542	8	INT/EXT	
MN150401	8	INT/EXT	
MN150402	8	INT/EXT	
MN158819	8	INT/EXT	
MN158281	8	INT/EXT	
MN158486	8	INT/EXT	
MN158418	8	INT/EXT	
MN1584532	8	INT/EXT	
MN158416	8	INT/EXT	
MN158614	8	INT/EXT	
MN158655/855	8	INT/EXT	
MN158851	8	INT/EXT	
MN15834A	8	INT/EXT	
MN158341	8	INT/EXT	
MN158412	8	INT/EXT	
MN158413A	8	INT/EXT	
MN158614	8	INT/EXT	
MN15151	8	INT/EXT	
MN15287	8	INT/EXT	
MN152810	8	INT/EXT	
MN152121	8	INT/EXT	
MN1700 Series	8×2	INT/EXT	

Table 5 cont.

	Bits	Clock	Max rate
Sanyo			
LC6520/22	4/8	INT/EXT	Clock/4
LC6530/32	4/8	INT/EXT	??
LC6538D	2×8/1×16	INT/EXT	??
LC6543/46	4/8	INT/EXT	Clock/4
LC6554	4/8	INT/EXT	Clock/4
LC6568	4/8	INT/EXT	Clock/4
LC65102/4	8	??	??
LC65204	8	??	??
LC66304/6/8	2×8/1×16	INT/EXT	Clock/8
LC66506/8	2×8/1×16	INT/EXT	Clock/8
LC66512/6	2×8/1×16	INT/EXT	Clock/8
LC67216	2×8	INT/EXT	Clock/8
LC5863/64	8	INT/EXT	Clock/2??
LC5866/8	8	INT/EXT	??
Seiko Epson			
SMC6235	8	INT/EXT	Clock/1
SMC623A	8	INT/EXT	Clock/1
SMC624x	8	INT/EXT	Clock/1
SMC6262	8 (mask option)	INT/EXT	Clock/1
SMC6266	2×8 (async)	EXT option	4800 baud
SMC6274	8	INT/EXT	Clock/1
SGS-Thomson			
ETx93xx	4	INT	Clock/4
ETx94xx	4	INT	Clock/4
Sharp			
SM55x	8	INT/EXT	Clock/4
SM57x	4/8	INT/EXT	Clock/4
SM5E4	8	INT/EXT	Clock/4
SM5Jx	8/4/2/1	INT/EXT	Clock/4
SM563	8	INT/EXT	Clock/4
Texas Instruments			
No serial I/O available.			
Toshiba			
TLCS-470A family	4/8	INT/EXT	Clock/64
TLCS-470 family	4/8	INT/EXT	Clock/64
TLCS-47 family:			
TMP47C200A/400A	4	INT/EXT	Clock/128
TMP47C210A/410A	4	INT/EXT	Clock/128
TMP47C212A/412A	4	INT/EXT	Clock/128
TMP47C220A/420A	4	INT/EXT	Clock/128
TMP47C221A/421A	4	INT/EXT	Clock/128
TMP47C423A	4	INT/EXT	Clock/128
TMP47C425	4	INT/EXT	fs/4
TMP47C231A	4/8	INT/EXT	Clock/64
TMP47C440A	4	INT/EXT	Clock/128
TMP47C441A	4	INT/EXT	Clock/128
TMP47C242A	4 o/p only	INT	Clock/128
TMP47C446A	4	INT/EXT	fs/4
TMP47C452A	4	INT/EXT	Clock/128
TMP47C453A	4	INT/EXT	Clock/128

Table 5 cont.

	Bits	Clock	Max rate
TMP47C456A	4	INT/EXT	fs/4
TMP47C460A	4	INT/EXT	Clock/128
TMP47C475A	4	INT/EXT	fs/4

Table 6 Parallel I/O capabilities

	Ports × bits	Comments
Fujitsu		
MB885x,/B,/F,/H	1×8 Out	PLA coded
MB8850x,/H	1×8 Out	PLA coded
Hitachi		
H400 series	1×8 Out	Table generated
Mitsubishi		
M50720	1×8 Out	
M50722	1×8 Out	
M50726	1×8 Out	
M50727	1×8 Out	
M50920	1×8 Out	
M34203M4	1×8 Out	
M34210M2	1×8 Out	
National Semiconductor		
COP2xxC	1×8 I/O	
COP31xC/L/CH	1×8 I/O	
COP32x/C/L	1×8 I/O	
COP340	2×8 I/O	
COP341	2×8 I/O	
COP342	2×8 I/O	
COP344C/L	1×8 I/O	
COP345C/L	1×8 I/O	
COP41xC/L/CH	1×8 I/O	
COP42x/C/L	1×8 I/O	
COP440	1×8 I/O	
COP441	2×8 I/O	
COP442	2×8 I/O	
COP444C/L	1×8 I/O	
COP445C/L	1×8 I/O	
NEC		
μCOM75 family (except μPD7556/66 & μPD7554/64)	1×8 I/O	
μPD7500x	2×8 I/O	
μPD751xx	3×8 I/O	
μPD752xx	1×8 I/O	
μPD75268	1×8 I/O	
μPD753xx	2×8 I/O	
μPD75328	2×8 I/O	
μPD755xx	2×8 I/O	

Table 6 cont.

	Ports × bits	Comments
OKI		
MSM6404/8	1×8 Out	Table generated
Panasonic		
MN1500 series		Any pair of 4 bit ports can be used together for single instruction 8 bit operation.
MN1700 series		Memory mapped I/O with byte transfer, also 10 bit wide table output, and external memory mode operation.
Sanyo		
LC66xxx	1×8 I/O	(also table o/p)
Seiko Epson		
Several parts with I/O capacity for byte-wide or greater transfer, but no instructions to output more than 4 bits at a time.		
SGS-Thomson		
ET9400 family	1×8 I/O	(also table output)
Sharp		
SM550	1×8 I/O	
	1×8 Out	
SM551/2	2×8 I/O	
	2×8 Out	
SM578/9	1×8 I/O	
SM5E4	2×8 I/O	
	2×8 Out	
SM5J5/6	1×8 I/O	
SM510/1/2	1×8 Out	(for keyboard drive)
SM563	1×8 I/O	(shared with some LCD drive)
Texas Instruments		
No parts with more than 4 bit I/O in a single instruction.		
Toshiba		
TLCS-470A Series	1×8 Out	Table generated
TLCS-470 Series	1×8 Out	Table generated
TLCS-47 Series	1×8 Out	Table generated

Table 7 DTMF drivers and receivers

Hitachi	
HD404608	Generator
HD404678	Receiver
NEC	
Generator under development.	
OKI	
MSM6352	Generator
Panasonic	
MN158614	Generator

Table 7 cont.

MN15834A	Generator
MN158341	Generator
MN158412	Generator
MN158413A	Generator
Texas Instruments	
TCM8305B/7B	Generator
TCM8306B/8B	Generator
Toshiba	
TMP47C850	Receiver
TMP47C852	Generator
TMP47C855	Generator
TMP47C857	Generator
TMP47C858	Generator
TMP47C407	Generator
TMP47C45x	Generator

Table 8 Phase locked loops

Fujitsu
MB8856x

Mitsubishi
M50440
M50442
M50443
M50445
M34350N6

NEC
μPD170xx

Panasonic
MN15287
MN152121

Table 9 Memory provision

	RAM	ROM
Fujitsu		
Max RAM 256×4, max ROM 8k×8.		
Hitachi		
HD404019	992×4	16k×10
HD404439	960×4	16k×10
HD404519	512×4	16k×10
HD404608	1184×4	8k×10
HD404719	960×4	16k×10
HD404720	576×4	16k×10 + 8k×10 (pattern ROM)
HD404729	576×4	16k×10
HD404808	1184×4	8k×10

Table 9 cont.

	RAM	ROM	
Mitsubishi			
Max RAM 256×4 with 4k×9 ROM, Max ROM 6k×9 with 192×4 RAM.			
National Semiconductor			
	32×4 to 160×4	0.5k×8 to 2k×8	
NEC			
μCOM75 family	64×4 to 256×4	1k×8 to 6k×8	
μCOM75x family	320×4 to 1k×4	4k×8 to 32k×8	
μPD75048	512×4	8064×8	1k×4 EEPROM
μPD75402	64×4	1920×8	
μPD170xx	336×4	8k×8 32k×8	Some have EEPROM
μPD17102	222×4	2k×16	
μPD17103/4	16×4	512×16	
OKI			
OLMS 50/60 series	62×4 to 1024×4	1024×14 to 4096×15	
OLMS 64/65 series	32×4 to 512×4	1024×8 to 8192×8	
Panasonic			
MN1500 series	16×4 to 512×4	0.5k×8 to 12k×8	
MN158614	256+1024×4		
MN1700 series	128+96×4 to 512+96×4	2k×10 to 12k×10	
Sanyo			
LC57xx	48×4 to 128×4	2k×8	
LC58xx	64×4 to 265×4	1k×15 to 8k×16	
LC65xx	32×4 to 448×4	512×8 to 8k×8	
LC66xxx	512×4	4k×8 to 16k×8	
LC67216A	608×4	16k×8	
Seiko Epson			
SMC62 family	80×4 to 1024×4	1k×12 to 8k×12	
SMC4202	128×4	2k×10	
SGS-Thomson			
	32×4 to 128×4	0.5k×8 to 2k×8	

Table 9 cont.

	RAM	ROM
Sharp		
	32×4 to 256×4	508×8 to 8192×9
Texas Instruments		
TCM8305B/6B	256×4	2048×8
TCM8307B/8B	256×4	4096×8
TSS40x	144×4	2048×9
TSS40x/4	240×4	4096×10
TSS400-S230	6×8-digit	2k×8 external EEPROM
Toshiba		
TLCS-42 family	32×4	512/1024×8
TLCS-47E family	128×4	1k/2k×8
TLCS-470A family	512/768×4	12k/16k×8
TLCS-470 family	256×4 to 1024×4	4k×8 to 8k×8
TLCS-47 family	128×4 to 768×4	2k×8 to 4k×8

Table 10 Computing power

	Minimum instruction time (microseconds)
Fujitsu	
MB88xxxH	1.5
MB8857x	1.5
Hitachi	
HMCS402AC	1.33
HMCS404AC	1.33
HMCS408AC	1.0
HD404019	1.0
HMCS412AC	1.0
HMCS414AC	1.0
HMCS424AC	1.0
HD4044x8	1.0
HD404708/9	1.0
HD404720	1.0
HD404808	1.0
HD404919	1.0
Mitsubishi	
M5072x	1.6
M5092x	1.6
M34203M4	1.6
National Semiconductor	
Minimum cycle time in COP400 family = 4 microseconds, 16μS for low voltage, low power operation.	
NEC	
μCOM75 family	2.44 to 10

Table 10 cont.

	Minimum instruction time (microseconds)
μCOM75x family	1
μPD171xx	2
OKI	
MSM505x	122
MSM6051	91
MSM6351/3	61
MSM6352	18
OLMS 64/65 series	1
MSM6501/12	19
Panasonic	
MN1500 series all	2 except:
MN1551A	1
MN1554	2.32
MN158281	1.91
MN158486	1.91
MN158416	4
MN158851	1
MN15834A	2.23
MN158341	2.23
MN158412/3A	2.23
MN158810	8
MN1700 series	1
MN1700A series	0.5
Sanyo	
LC57xx	122
LC5732H	10
LC5733H/34H	18
LC58xx	244/122
LC5851N	60/40
LC586xH	2
LC6512A/13A	1.3
LC6514B/15B	3.5
LC65xxC	3
LC65xxH	0.92
LC65102A/104A/204A	0.65
LC66xxx	0.92
LC67216A	0.65
Seiko Epson	
SMC621x	153/11
SMC6232/33/35	153/10
SMC6237/3A	153
SMC6244/48	153/2
SMC6246	153/5
SMC6251	153
SMC6262	153/5
SMC6266	130/5
SMC6274	153/10
SMC6281	153
SMC6282	153/5
SMC4202	122

Table 10 cont.

	Minimum instruction time (microseconds)
SGS-Thomson	
ET9400 family	4 (5 volt operation) 16 (low voltage, low power operation)
Sharp	
SM590/1/2	1
SM550/1/2	1.6
SM578/9	2
SM5E4	1.6
SM5J6	2
SM5J5	2.5
SM530/1	91.5
SM500	61
SM5K1	5
SM4A	61
SM510/1/2	61
SM563	2
Texas Instruments	
TCM8305B/6B	23
TCM8307B/8B	23
TSS400 family	5.5
Toshiba	
TMP42xx	2.5
TMP42Cxx	1.0
TLCS-47E family	1.9
TLCS-470A family	1.3
TLCS-470 family	1.3
TLCS-47 family	1.9

Table 11 Special facilities

Fujitsu	
MB885xx	Optional watch-dog
Hitachi	
HD404302/308	Watch-dog
HD404408/418	Watch-dog
HD404439	Watch-dog
HD404608/808	Watch-dog
HD404719	Watch-dog
Mitsubishi	
M50725/925	4 timers
M50726/7	4 timers
M34201M4	Remote control transmit
National Semiconductor	
Extended temperature range parts available	
NEC	
μPD7519/19H/16H	8243 port expander I/F

Table 11 cont.

μPD7527A/37A, 28A/38A	Zero crossing detection
μPD751xx	Watch-dog timer
μCOM75 family	Bit sequential buffer
μPD17102	Zero crossing detection
μPD17203	Learning remote control
OKI	
OLMS 50/60 series	Buzzer drive (some parts with melody)
MSM5056	Solar powered
MSM635x	Watch-dog timer
Panasonic	
MN158381	Watch-dog timer
MN158417	Remote control transmitter
MN158819	Home bus I/F, watch-dog timer
MN158682/882	Watch-dog timer, buzzer drive
MN1584532	Remote control transmitter
MN158655/855	Remote control transmitter, buzzer driver
MN158851	Buzzer driver
MN15151	Remote control rx, TV tuner facilities
MN15287	TV tuner facilities
MN152810	TV tuner facilities
MN152121	TV tuner facilities
MN158810	Remote control transmitter
Sanyo	
LC57xx/58xx	Melody functions, input debounce ('chattering eliminator'), special chronograph functions, reset from input port, watch-dog timers
LC5734	Remote control carrier modulator
LC6538	Automatic key scan
LC66xxx	Watch-dog timer option
LC67216A	Remote control receiver
Seiko Epson	
SMC62 family and SMC4202 all have supply voltage detection (SVD)	
SMC6215/1A	Remote control O/P
SMC623x	Stop-watch, watch-dog
SMC624x	Sound generator, watch-dog
SMC6251	2×buzzer O/P
SMC626x	Watch-dog
SMC6274	Built in operational amplifier
SMC628x	Melody generator
SGS-Thomson	No special facilities
Sharp	
SM5J5	40 volt, 80 mA output
SM530/1	Melody generator
SM511/2	Melody generator
Texas Instruments	
TCM830xB range	Designed for telephone applications Melody (ringing) generator
TSS400 family	Designed for sensor signal processing
TSS400-S230	'TSS400 Standard' preprogrammed with interpreter for program held in external EEPROM
Toshiba	
TLCS-470A family	Remote control preprocessors

TMP47C434/634	Remote control preprocessors
TMP47C662/862	2×12 bit pulse generator
TMP47C670/870	Keyscan
TMP47C440/1/6	Watch-dog timer
TMP47C242	Watch-dog timer
TMP47C456/7	Watch-dog timer
TMP47C475	Watch-dog timer

Table 12 Packaging options

Fujitsu Most parts available in DIP, shrink DIP, QFP and/or SOP.

MB8851/5&/H Available with piggyback EPROM sockets. No OTP versions or bare chips.

MB885xx families available with piggyback EPROM sockets, and most parts have OTP versions. No bare chips.

Hitachi All mask programmed parts available in quad flat pack, thin quad flat pack (1.0 mm thick) being introduced. 42 and 64 pin dual-in-line packs for smaller devices. No bare chips.

User programmable equivalents in dual-in-line (EPROM or OTP) and flat pack (OTP only). Piggyback EPROM equivalents available for general purpose types.

Mitsubishi Packaging options include DIP, shrink DIP, SOP and QFP, with piggyback versions in DIP, shrink DIP and QFP. No bare chips.

National Semiconductor COP400 family available in a variety of dual-in-line and surface mount packages. ROMless devices mainly use separate ROM, but some use piggyback. No bare chips.

NEC NEC parts are available in a variety of packaging options, according to the requirements of the particular device. Dual in line parts include windowed and piggyback varieties. Flat packs are commonly used for parts with more than 40 leads, and chip carriers are also available for some parts. Bare chips are not normally supplied.

OKI Many OKI parts are available in naked die form, as well as packaged, mainly in DIP and quad flatpack. Piggyback parts are available for the OLMS-64/65 series.

Panasonic

MN1500 series DIP, SO, and QFP as appropriate, with piggyback versions for all but the MN158453

MN1700 series SDIP/QFP, with piggyback versions. Built-in EPROM and OTP available for MN17P58.

Sanyo

LC57xx, LC58xx (LCD drive parts) all available in bare chip and quad flat pack form. No user programmable parts.

Other parts in DIP, quad flat pack, and some in surface mount dual in line packages. Piggyback parts available for most non-LCD parts, plus some windowed EPROM, and some OTP parts.

Seiko Epson

SMC62 family and SMC4202 Bare chips and quad flat packs. No user programmable parts.

Table 12 cont.

SGS-Thomson

ET9400 family	Dual in line for all parts, plus small outline surface mount packages, and leadless chip carriers for selected parts. No bare chips. ROMless parts and piggyback versions not available from SGS-Thomson. Use National Semiconductor parts.

Sharp

Controller series and VFD driver series parts are mostly available in DIP or QFP according to size. Some parts available in both styles. LCD driver series available only in QFP. No bare chips. Not well supported by user programmable parts. One OTP available.

Texas Instruments

TCM830xB	28/40 pin DIP or 28/44 pin PLCC. No user programmable parts.
TSS400	28/40 pin DIP or 44 pin PLCC. TSS 400 Standard is user programmable.

Toshiba

TLCS-42 family	DIP/SDIP/SOP, piggyback board.
TLCS-47E family	DIP/SDIP/SOP, OTP available.
Other families	SDIP/QFP, well covered by piggyback and OTP versions.

Table 13 Second sourced parts

National Semiconductor Most parts second sourced by SGS-Thomson.
Part numbers translate quite easily from National COP numbers to SGS-Thomson ET9 numbers. For example:

Nat Semi no	SGS-T no	
COP420	ET9420	
COP310L	ETL9310	
COP411C	ETC9411	etc

but note that there are exceptions to the rule:

COP424C	ETC9420
COP425C	ETC9421
COP426C	ETC9422
COP324C	ETC9320
COP325C	ETC9321
COP326C	ETC9322

There is no second source for the National COP2xxx range (Military Temperature versions), or for the COP413CH/COP313CH.

4 Hardware design I – Communications, interfacing, external memory, power supply considerations and power saving techniques

In many respects the 4 bit microcontroller is similar to any other component in a digital circuit design. It needs a power supply with appropriate levels of decoupling to minimise noise on the power lines just as other digital circuits do. It also has to draw information in from other parts of the system in order to process it, and it has to be able to pass the results on through some form of output, be it a parallel data bus, a serial data link, operator display or other form of communication. In this it is no different from many other system components. However there is such a wide range of applications for 4 bit microcontrollers, and such a wide range of power requirements, circuit possibilities, etc, that it is worthwhile examining the hardware design aspects in some detail, in an attempt to identify the special design techniques that need to be adopted when working with these parts. This chapter and the next will therefore be devoted to hardware design, and cover such topics as power requirements, clocks and oscillators, communications, interfacing, external memories, power saving techniques, input and output, displays, keypads, bleepers and other peripheral circuits.

In general the 4 bit microcontroller is an ASIC (application specific integrated circuit) mask programmed for each circuit or system in which it is used. Some devices are available for use with EPROM, or are OTP allowing the user to program the device rather than going to the expense of mask programming. But the majority of devices are mask programmed, and are therefore applied where there is sufficient volume to justify the cost of masking, and enjoy the benefits of the low per unit cost thereafter. The microcontroller will be expected to mop up large sections of a system, making possible smaller, lighter, less power consuming or more versatile implementations than would be possible without it. The 4 bit microcontroller is, in general, cheaper and less power consuming than its larger and more powerful counterparts which work with longer wordlengths. The aim in making use of a 4 bit microcontroller in any circuit is to minimise cost and maximise functionality. This is achieved by reducing parts counts, PCB area and power requirements.

Power supplies

4 bit microcontrollers can be divided into two broad groups as far as their power requirements are concerned. In one group are the devices designed to operate from a 5 volt supply, probably derived from a mains supply or high capacity battery such as in a motor vehicle. Devices in this group will typically consume a few milliamps of supply current. In the other group are the parts which operate from lower voltages, typically 3 or 1.5 volts, and which are designed to be used in very low power equipment, powered from a single small cell, which is expected to have a life of a year or more. This group of parts will be operating in the microamp region most of the time, though for short periods may require substantially more supply current, for example during audio output.

The power requirements of the first group are more akin to those of conventional digital electronics. The majority of parts on the market are CMOS technology, though there are some N-channel parts about. In general the N-channel parts require tighter supply regulation than the CMOS parts, and it is important to study the manufacturer's data on the chosen part to make sure that the power supply design is adequate. It is also important to remember in designing the power supply to include the current controlled by the microcontroller, for instance to drive a display, or perhaps supply control current to some type of switch. Very often the controlled currents will be from a different power source from the main logic supply, as for example when an elevated voltage is required by the display being used. Also very important is the environment in which the circuit has to operate. Often controllers are required to work where there are large inductive loads being switched, motors, fans, pumps and the like, which contribute to a very noisy mains supply. The subject of power supply design, and how to minimise the effects of such external interference, is well beyond the scope of this book. However, the fact remains that it is sometimes extremely difficult to prevent these disturbances from reaching the low voltage logic power supply. Careful steps must therefore be taken in system design, both hardware and software, to ensure system integrity in the event of major power supply disturbances. It is likely that all sorts of operating parameters will be held in the internal memory of the microcontroller, which are likely to 'evaporate' in a surprisingly short time if the power supply drops out of the specified range. Steps to ensure that the system can cope with this can be built into the software, and this is covered in Chapter 7. More serious perhaps is the over-voltage situation, as this can cause breakdown and permanent malfunction in a microcontroller, so must be avoided at all costs. Fortunately there are a number of quick acting transient absorbing devices on the market which can be built into power supply design if this is likely to be a problem in a given environment.

Another important environment where 4 bit microcontrollers are to be found is in automotive applications. The automotive environment is a particularly difficult one, with extremes of temperature, all sorts of dirt and unpleasant substances, and a difficult electrical regime. Again, a detailed treatment of this subject is beyond the scope of this book, and the reader is referred to publications on the topic by the International Standardization Organisation and other specialist publications. It is probably sufficient to say here that the use of a voltage regulator specially designed for automotive applications is recommended.

At the other end of the scale are the very low power devices, designed for applications such as wristwatches and pocket sized equipment which will be expected to operate for a year or so from a miniature cell, offering only a few tens of milliampere hours of energy supply. These parts are invariably CMOS, that being the only technology currently capable of offering such performance. There are a number of points to bear in mind when designing for minimum power consumption. All circuits contain a certain amount of capacitance, and the action of digital circuits consists of charging some capacitance from the supply, and then when the circuit switches, allowing the charge to decay. The power consumed by this process is probably the major source of power consumption in a CMOS circuit, and is affected by the amount of capacitance, the voltage, and the frequency at which it is shifted. In fact the current consumed is given by the simple relationship:

$$I = V \times C \times f$$

where V is voltage in volts, C is capacitance in farads, and f is frequency in hertz. It can be seen from this that a simple oscillator operating at 32 768 Hz, the usual value for circuits required to keep accurate time, with a capacitance of 20 pF, which is about the minimum practical value for stable oscillation, will consume about 2 μA when operating at 3 volts. When trying to squeeze maximum life out of a small capacity cell 2 μA is not an insignificant current. It pays therefore to minimise circuit capacitance, and operate at the lowest possible speed. Lowering the voltage obviously reduces the power also, but the choice is limited by available cells, and the lower voltage units usually have a lower capacity per unit volume, so there is little to be gained there. Concerning speed of operation, much depends on the application, and how demanding it is of the computing power of the controller used. In most applications there are times when the processor is required to operate flat out, while most of the time it is fairly dormant. This suggests that some kind of variable speed clock should be employed, but in fact it is usually achieved by making use of STOP and HALT states in the microcontroller. Power saving techniques under these circumstances are therefore mainly confined to the software – see Chapter 7. It is also important not to violate the manufacturer's specification. Many devices have a minimum clock frequency as dynamic storage techniques are used for some of the internal registers, and correct operation of the part cannot be guaranteed below the stated minimum clock frequency. There are some parts though where dynamic storage is not used, and the clock can be stopped by some external means when the controller is not needed to operate. This should put the device into a minimum power configuration, so watch out for terms like 'fully static' in data sheets. There is also a limit to the extent to which the system designer can minimise capacitance. Much of the capacitance is internal to the controller and is beyond the control of the user. Predicting current consumption can be difficult, but curves published by manufacturers for various clock frequencies and operating voltages will usually give some idea of typical values. Clearly it is important to minimise all external load capacitances wherever possible.

There are two other considerations when predicting and minimising power consumption. The first is the quiescent current required by the controller. This is entirely dependent on the internal design of the circuit, and is mainly due to leakage currents. The manufacturers' data sheets can be the only guide here, but note that the leakage current is usually a thermal effect, and increases with increasing temperature. So even a circuit which is stopped, or operating for a very small percentage of the time may consume alarmingly high currents at elevated temperatures. For example, the data sheet for one device gives a typical quiescent current of 0.3 μA, but specifies a maximum of 10 μA over the operating temperature range.

The second point to watch is switching speed. CMOS circuits operate by switching on either a P-channel device connected to the positive supply or an N-channel device connected to the negative supply. During switching it is possible for both devices to be on for a moment while the switching signal passes through the threshold voltages of the two transistors. Clearly the more quickly this transitional zone is traversed, the less charge will be able to drain away. It is important therefore to ensure that slowly changing voltages are not applied to inputs which are not specifically designed to accept such signals. Most controllers will have some inputs which include a schmitt trigger circuit which has some built in hysteresis, and can cope quite successfully with slowly changing voltages. If an application does include signals of this nature it is important that appropriate inputs are used. Incidentally, this current which flows during switching between states in a CMOS gate also contributes to the increase in power with increasing frequency effect, though the capacitive loading is usually the dominant part, and the transient current gives an apparent increase in internal capacitance. It is also important to make sure that no unused inputs are left 'floating', unconnected to either supply rail. If they are they can easily drift into the state where both P and N channel transistors switch on, and even if this is only partial switch on, it can cause comparatively large amounts of battery power to go to waste.

Batteries

The subject of batteries is a huge one, so remarks here will be confined to a few brief comments about the kinds of small single cells that are available for use with low power microcontrollers. The silver oxide cell, which gives a voltage of 1.5 volts, is well established as a power source for watches, calculators, cameras, etc, areas where 4 bit microcontrollers can be used to increase functionality of equipment, and often make it simpler to use. The reason that a number of manufacturers offer controllers which operate at 1.5 volt is the availability of this power source. However, a 3 volt source is much more versatile, and allows a greater choice of controllers, as well as increasing the possibilities in peripheral devices such as extra memory, or special sensing amplifiers. The advent of the lithium cell has given the equipment designer a handy power source with a high energy density,and it is now well established in the watch and calculator field. Progress is constantly being made, and cells which pack more ampere-hours into the same size of can are appearing all the time. The designer must therefore scour the catalogues of the battery manufacturers to discover the power source that fits the system requirements.

Clocks and oscillators

The subject of clocks and oscillators was touched on briefly above, in discussing the effect of oscillator design on power consumption. The conclusion was that clocks should be as slow as possible, and oscillators use the minimum possible capacitance in order to minimise system power use. This is perhaps a rather simplistic approach, as there are other considerations to bear in mind, not least of which is the computing load which the microcontroller has to perform. A slow clock means slow processing, and inevitably a compromise has to be made between power consumption and processing speed. All microcontrollers require a clock of some sort, some require more than one. The manufacturer's aim in this area is to minimise the number of external components required to implement a suitable oscillator, and at the same time give the user a reasonable control over choice of operating speeds, clock stability, etc. The choice is usually between a crystal, a ceramic resonator, or a simple RC circuit, and the decision as to which of these to use depends on the application. If accurate timing is required (to ± a few parts per million) then a crystal will have to be used. A great many parts are designed to be used with the standard watch crystal with a resonant frequency of 32 768 Hz, to allow accurate time of day functions, and also tone generation, for example in telephone DTMF diallers. As 32 kHz is rather slow as a processor clock, some parts are made which allow this crystal to be used with some timer/counter registers, while employing a higher frequency clock to run the processor. This higher frequency clock can then be a low accuracy, low cost, RC oscillator. This has the advantage of allowing the processor to operate at a higher speed and at the same time keeping the convenience of the 32 kHz clock for timing purposes. There are various arrangements of dual clock systems available. In some parts the processor, along with the high speed clock, can be stopped, while the 32 kHz clock continues to run, allowing the timer/counter to continue and restart the processor by using an interrupt when it times out. This is a handy power saving technique, and allows the system to have sufficient processing power when needed, but to operate at low power when it can. In some parts the 32 kHz clock is able to take over from the high speed clock to run all, or selected parts, of the microcontroller, allowing some processing to be carried out in the lower power mode.

Ceramic resonators offer a lower cost alternative to crystals, but with lower accuracy, typically ± 0.5%. This may be sufficiently accurate for many applications, including tone generators, speech synthesis applications and the like. The final option, of an external resistor and capacitor to determine oscillator frequency is clearly the most economic, but frequency is only as accurate as the components used, and this method should only be used where there is no need for accurate timing from the processor clock. It will be quite adequate in a large number of applications.

The detail of oscillator design has been taken care of by the designers of the microcontroller, so all the user has to worry about is selecting the correct external components to suit a given application, and making sure that the manufacturer's data sheet recommendations are followed carefully. One point to note when choosing an oscillator is the difference in start-up performance. If the clock is stopped for some reason, a crystal or resonator circuit will take a few cycles to establish stable oscillation, whereas an R C circuit will immediately start oscillating in the required way once power is applied or inhibit signal removed.

Communications

Microcontrollers are often used in systems where they are required to communicate with other devices. Indeed in many applications this is their main task. The communications covered here are those involving other electronic devices; communication with an operator via keypads and displays is covered in Chapter 5. In this section the subjects under consideration are the available media for inter-device communication, and appropriate data rates, coding techniques, and signal levels for the application, particularly considering the distance over which communication is being made.

Devices can be directly coupled by wire, or can use some other intermediate medium such as radio, ultrasonics, or light.

Electrical communications

To connect two circuit elements together by wire or printed circuit track is the most natural and obvious method of communicating between them. Whether signalling is at logic voltage levels, or at elevated voltages, using single ended or differential transmission, operating in voltage or current mode are all choices to be made according to the application. Design choices here also include whether to use parallel or serial data transfer, what rate of transfer to use, and what wordlength is most appropriate.

Parallel data transfer

For high speed data transfer over short distances the parallel method is the best and simplest to implement. It is applied almost universally as the primary method of data transfer within systems controlled by microprocessors with longer data wordlengths. 4 bit microcontrollers operating under the control of such may have to operate on a parallel data bus, see under Data buses below. One to one parallel links can be reasonably simple to implement in 4 bit systems, especially if the chosen microcontroller has a strobe facility. Some devices include this, where an output line can be designated for use as a strobe line, and automatically gives a pulse out when another output port is loaded. Output ports are usually only 4 bits wide, but they are mostly latched so that the data remains on the port as long as required, so that data can be output in two nibbles even if the receiver requires byte-wide transfer. A strobe signal can be activated after the second nibble of data has been loaded on to the output port.

In some parts there are instructions available for loading data onto two 4 bit output ports simultaneously, giving the effect of having an 8 bit output port. Reception of 8 bit data in this way is not so simple, as it requires more time for the microcontroller to poll two ports for the data input. It will often be possible to connect the incoming strobe line to an external interrupt input, so that data can be gathered in an interrupt routine. If timing is critical, and the data is not going to be available for long enough to use this method, look out for a microcontroller which allows two ports to be coupled together and used as an 8 bit input port. Parallel links are not usually used over distances greater than a metre or two, and if the signals are going off the board on which they originate they will often need to be buffered. The capacitive loading introduced by long lines can slow the signal rise times and hence limit the data rate. Buffer circuits can introduce skew, and delays, and hence also slow the data rate. Very often a more complex handshaking procedure than the simple strobe signal will be required. It is very common to employ an acknowledge signal as well - generated by the receiver this tells the transmitter when it is safe to change the information on the data lines. It is found in such systems as parallel printer interfaces.

Serial data transfer

Parallel data transfer has its disadvantages, particularly for 4 bit microcontroller systems where the number of I/O port pins required by a parallel link can be excessive, as often the I/O is at a premium. An alternative frequently employed is the serial link, and the majority of 4 bit microcontrollers have provision for serial communications. Although the basic working unit of the 4 bit microcontroller is a 4 bit word (nibble), provision is made in many devices to handle 8 bit words (bytes) for serial data transfer. Serial transfer over short distances is usually of the clocked, or synchronous, variety, using three lines, data in, data out and clock, operating at normal logic levels. For serial output, the data to be sent will usually be in the RAM. The 8 bit shift register of the serial interface provided on the microcontroller chip is loaded from the RAM, and on a 'start transfer' command, data is clocked out of the shift register across the serial link. How easy it is to load the data into the serial interface depends on the microcontroller. In some devices the RAM address is specified, and two nibbles are automatically transferred from adjacent RAM locations. In other parts the data may have to be transferred via the accumulator or other register, possibly in two separate operations. The clock signal can usually be external or internal, allowing control of the transfer to be with either the sender or the receiver. The device generating the clock signal for the data transfer is usually termed the master, and the device operating from an external clock is the slave. Serial input is similar, data being clocked into the shift register from the external source, and when the register is full, the data is read out into two memory locations, or a register and a memory, or sometimes a pair of registers. Often there is an interrupt signal associated with the serial interface, so that an interrupt is generated when new data is received, or when the buffer is empty and ready to send more. On some microcontrollers it is possible to send and receive simultaneously, as the one clock clocks data into one end of the shift register while it is clocking data out of the other end. There is often provision also to specify which way the data goes, most significant or least significant bit first, i.e. the direction of the shift register.

Serial links are usually quite simple to handle in software, and once loaded need no attention until the interrupt occurs for the next byte to be loaded. With a short electrical link of this nature there is usually no need for much in the way of protocol. Such serial links operate without start bits and stop bits, and it is usually quite simple to add any extra signal lines to indicate when data can be transferred if this is necessary. It may be that the simple synchronous interface is not suitable for a given application, in which case an asynchronous link could be considered. This may be necessary because of the distance over which the link has to operate, or it may be a requirement of the device with which the microcontroller is communicating. Modems and printers for example often operate using asynchronous serial communications. In asynchronous communications one data line is used for transmission in each direction, the voltage on these lines being referred to a third, zero volts, line. The transmitting circuit normally holds the signal line at one voltage level (often the high voltage), switching to the other at the start of transmission of a byte of data. The data is held at this level for the duration of one bit time, this being the 'start bit'. The data bits follow at a predetermined rate, and after all the data bits have been sent the line is returned to the original voltage level for at least one bit time (the 'stop bit'). Some systems specify 1.5 or 2 stop bits. The data rate in bits per second at which data is transferred is commonly known as the baud rate. Using this method the data transfer is synchronised on every start bit, and the receiver has to measure the time after detecting the start bit to determine when the signal line should be sampled for each of the data bits. Parity is often used as a check on the integrity of the data, and there are a number of other error detection and correction techniques employed for long distance data communication.

As far as the signal voltage is concerned, there are several transmission standards for data transfer over long links. The commonest is probably the RS232 which uses signal levels of $\pm$ 9 to 12 volts, using a single line in each direction referred to a third, ground, line. Also commonly used is RS422 which uses a differential system, and hence is less susceptible to interference, and more appropriate for noisy environments, such as in industrial control. Also becoming increasingly popular in consumer applications is the domestic digital bus (D2B), which is another differential system, designed to link a number of devices. There are many interface circuits available to translate logic level signals to the appropriate standard, some of which will also generate the extra voltages required from the logic supply. In the case of interface standards like D2B the interface circuits will also carry out protocol handling, making the applications programming for the microcontroller even simpler.

From the hardware point of view it is a simple matter to connect a few of the microcontroller I/O pins to the interface circuit – the complications come in the software to generate signals to the required timing standard. There are a number of microcontrollers available now which integrate a full hardware UART (universal asynchronous receive/transmit) circuit, sometimes two of them. This means that the program just has to set up hardware registers specifying operating parameters such as baud rate, parity, number of stop bits, etc, and then just pass data to the UART port, or receive data from it. In other microcontrollers, where there is no such provision, there is often available a standard software library which includes code to implement a UART in software, and provide all the facilities that would be expected from a hardware version. There are some examples of software implementing asynchronous serial communication in Chapter 8.

Data buses

A data bus is a channel for connecting a number of digital circuits so that they can communicate with one another. In the discussion of communications above the assumption has always been that there is a one to one communication process, between the microcontroller and another device. Often it is useful to be able to use one wire, or group of wires, to serve as the medium for communication between a microcontroller and several other devices. There are a number of ways of implementing such a data bus. One way, touched on briefly above, is to use a special purpose bus chip between the microcontroller and the bus. This is how the D2B, and a number of other bus protocols, are operated. Of more interest to the developer of a small system are techniques for implementing a bus structure directly from the microcontroller, without the need for special interface chips.

There are basically two types of data bus, the serial bus and the parallel bus. The parallel bus is the usual means of communication within 8 and 16 bit systems, where microprocessors require external memory, and often the I/O is mapped into the memory address space, or shares the data bus in some way. The parallel bus has to carry address information as well as data, and sometimes the two are time division multiplexed on to the same set of lines, while in other systems the bus has to be wide enough to take address and data separately. Because of the profusion of 8 and 16 bit systems, there are a great many parts on the market which provide peripheral functions and are designed to interface to a processor via a parallel bus. The functions which on larger systems require a data bus are integrated on to the 4 bit microcontroller, and so the data bus is a completely internal affair. Implementing a parallel bus with a 4 bit microcontroller can be clumsy, and use up a large proportion of the I/O available. They are not really designed with a parallel bus in mind, and aim to integrate the functions that on larger systems require a parallel bus structure. However, if it is necessary to operate a parallel bus on a 4 bit microcontroller, the essential ingredient is the I/O port which can be put into high impedance mode while another device on the bus is transmitting data to the microcontroller, and can be used to drive the bus when it is required to transmit data to the peripheral devices or supervising microprocessor. Operating a bus which multiplexes address and data can put a high software overhead on to the system, and make it operate rather more slowly than a parallel bus is designed to work. A separate address section will use up even more I/O port lines.

The serial bus is an alternative structure which has been developed with much more of an eye to the 4 bit microcontroller system. There are several protocols, such as S-Bus and I2CBus, for which peripheral devices have been developed. The usual idea of the serial bus is to have a clock line and one or two data lines. There is a start signal of some sort generated by the microcontroller, and then the microcontroller clocks the address on to the bus, together with a read/write bit. The unit on the bus whose address matches the address sent will then respond by accepting data if the microcontroller is writing, or sending data if the microcontroller is reading. A common use for this type of bus is to have an external non-volatile memory for storage of vital parameters which must not be lost when the power goes off. A number of manufacturers make EEPROMs which conform to serial bus standards. There are other peripheral functions available for use with a serial bus, such as audio signal processing devices for use in entertainment systems, and other functions which would not normally be integrated onto a microcontroller chip because of their rather specialist nature, or special semiconductor processing requirements. Such a serial bus can often be implemented relatively simply using the microcontroller's own serial port, which will usually have the clock signal, and input and output lines, needed for the serial port, and will also have instructions to allow data to be loaded in parallel into the serial port, and to allow the actual serial transfer to take place without software intervention. Some microcontrollers have special facilities built into their serial ports to simplify their use in serial bus applications. For example an address comparator is sometimes provided to simplify operations when the microcontroller is acting as a slave unit on the bus. The unit's address is loaded into the appropriate register, and the serial interface performs a comparison with incoming data. If the address is matched an interrupt is generated.

Optical links

There are applications where it is necessary for two units to communicate but making an electrical connection between them is inappropriate for some reason. For instance, a hand-held data store might be linked to a personal computer in this way, the data store containing a light source beneath a transparent window fits on to an interface cradle containing the receiver. When the data store is placed in the cradle the sender and receiver line up and are closely coupled, behaving in a similar way to an opto-isolator.

It is simple to design the case to resist the ingress of dirt and moisture, and there is no danger of damage from static discharge or other high voltage input as there might be if there were electrical contacts on the outside of the case. Another advantage of this type of short optical link is the high immunity to electrical interference, which can be crucial in some applications. The link is usually highly directional, and care has to be taken with the mechanical design to ensure proper alignment of transmitter and receiver. There is also a choice to be made as to what frequency of light to employ – visible or infra red. On the whole, infra red offers better opportunities for rejection of interference from other sources of light, usually by fitting an infra red filter over the receiver.

Sending data in this way is usually very simple, the transmitter consisting of one LED. Many microcontrollers have outputs capable of driving LEDs directly, and even if the outputs do not have sufficient current drive capability, a single transistor will be all that is required to give the necessary current gain. To keep the cost and power to a minimum the interface will probably be limited to one line, suggesting a serial link with a fixed bit rate. The microcontroller's own serial port will probably be unsuitable for such a link because of the need to add a start bit to the data, and because there will not be enough control over data rate and pulse length. If power is at a premium the pulses emitted by the light source can be made very short. If the instruction time of the microcontroller is too long it may be necessary to AC couple the signal between microcontroller and LED. Panasonic have an instruction for their MN1500 series which could be useful here – it reverses the output data on the specified port bit for a quarter of a machine cycle, giving the possibility for a 500 nS pulse. In the simplest coding technique a pulse is output for a 1 and no pulse is output for a 0, using a fixed data rate. If the main clock which runs the microcontroller is a simple RC clock it may also be necessary to devise some simple protocol to establish the data rate, as the clock rate will vary from device to device. Self-clocking codes can also be used, for instance a single pulse could be used to represent a 0 and a double pulse to represent a 1. This method uses a little more power than the simpler pulse/no pulse scheme, but it does avoid the problems of small variations in the data rate.

One important class of applications for 4 bit microcontrollers is the infra red remote controller, a hand-held battery operated device which sends out coded signals to control an increasing range of, in the main, domestic devices. The manufacturers have recognized the importance of this application and have produced families of devices specifically geared to it. This is a special example of optical communications, and requires a fair degree of sophistication from the software, the protocols and coding having to be quite complex to ensure correct response.

Data reception via an optical link is more complex than transmission. Some amplification will be required to get logic level signals. There is also a power penalty - the transmitter need only be active for a few microseconds or so while a pulse is being sent, but the receiver has to be on continuously whenever a pulse is expected, and will probably require an order of magnitude more power than transmission. In low power, battery operated equipment this means that there will have to be some means of switching off the receiver when it is not in use, otherwise an unacceptably high current drain would result.

High data rates can be achieved with optical links. The limit on the data rate will probably be the microcontroller, not the link.

To communicate optically over anything more than about ten metres it is necessary to employ an optical fibre transmission link. There are a number of proprietary systems available allowing for distances up to about 10 kilometres, and transmission rates of 10 megabits/second or higher. The interface between such a system and the microcontroller is at logic levels, and the transmitting and receiving circuits are usually packaged in a housing which facilitates the connection to the optical fibre. There may be complications if it is required to multiplex the fibre, but generally speaking the link can be treated in a similar way to a wire link. The controlling software will have to cope with such items as error checking and correction.

Radio links

The term 'radio links' is used here for communication between two units separated by a small physical distance using electromagnetic radiation, using very low power signals. Strictly speaking this also covers the region of inductive coupling, the difference between the two being the frequency of the carrier signal employed. Below 150 kHz the communication is regarded as inductive. Whatever the frequency employed, one way of looking at the system is to regard it as a transformer where the primary and secondary windings have been separated, so that the coupling between them is very loose. To make use of such a link some frequency selective circuitry has to be designed, and a carrier frequency chosen. The simplest coding technique is to use pulses of carrier in a similar way to the pulses of light in the optical link, which is quite efficient in terms of power requirements. This simple method can operate quite satisfactorily with quite low Q tuned circuits, but it is susceptible to interference from other electronic equipment. With a power of 100 microwatts or so it is possible to achieve a range of about a metre quite easily with quite a low frequency carrier, say 10 kHz. Some systems may demand more sophisticated schemes using a modulated carrier, frequency modulation being the commonest. One advantage is that there is always a carrier present and so it is possible to detect the out of range situation. The choice of frequency depends partly on the data rate required – too low a carrier frequency will restrict the data rate unnecessarily. It is also important to ensure that the chosen system does not fall foul of the regulations governing the emission of electromagnetic radiation in any country where it is expected to be able to market the product. Implementing such a link requires quite a bit of circuitry outside the microcontroller, including an aerial coil. If two way transfer is required, the coil should be capable of acting as both the transmitting and receiving element.

A radio or inductive link is a useful way of communicating if the communicating parts cannot be brought into contact, and a certain amount of freedom of movement is required. It offers various advantages in terms of being able to seal units completely from the environment, and avoid external electrical connections. There are one or two adverse points to note, particularly EMI limits mentioned above, and problems of shielding by metal objects, or even by water. Aerial coils are directional in nature and so the range will vary with position, and there is also the possibility of interference, not least from another similar unit operating nearby.

Short sonic/ultrasonic links

Several manufacturers now produce parts with built-in DTMF generators and decoders, aimed at the telecommunications market. The tones produced and sent down the phone line, or received over the phone are probably never anything but electrical signals in normal use, but there remains the possibility of generating sound signals, as in tone pads and telephone number stores which use DTMF to dial a number acoustically, using the telephone mouthpiece as a transducer. There is no reason why these sonic links should be limited to use in telephone related systems, but that is an obvious application area. DTMF tones stand poised to become a standard for all sorts of control applications, particularly with home and office equipment, whether they are coupled to the telephone or not. The main drawback to DTMF is the data rate, which limits its use to short control command applications rather than extensive data transfer. Sonic communication is already in use with data transfer as well, again in connection with the telephone, where the acoustic coupler has been working alongside electrically connected modems for many years. The applications for this kind of sonic communication are clearly limited, but there may well be situations where it is appropriate.

Ultrasonic communication is another rather specialised technique, which has its application in areas where other forms of communication are difficult, for example, under water. Ultrasonics suffer from high attenuation when operating through air, the attenuation increasing with carrier frequency, but they do well in denser media, provided the transducers can be coupled efficiently to the transmission medium at both sender and receiver. Discontinuities in the transmission medium also present problems. At any sudden change of refractive index there will be reflections generated, and hence signal attenuation. There are techniques of selecting frequency so that the wavelength of the signal in a thin barrier is such that reflections reinforce the signal rather than interfere with it, but that is beyond the scope of this book.

It should be noted that transmission of ultrasonic signals usually requires use of elevated voltages to achieve reasonable range, and reception requires high gain amplifiers, resulting in a considerable amount of circuitry external to the microcontroller.

I/O expansion

It may be that a particular microcontroller has all the correct attributes for a certain system, but just not quite enough I/O. The solution may be to expand the I/O capabilities with an extra chip such as the industry standard 8243 chip, which will give four extra 4 bit I/O ports at the expense of one 4 bit I/O port plus one control line. Some microcontrollers make provision for interfacing to the 8243 by providing a line which performs the functions required of the control line. Even if this specific provision is not made it is still possible to use an 8243, particularly if there is single bit manipulation available on an output port.

External memory

The aim of the 4 bit microcontroller is that as much as possible should be integrated on the microcontroller chip. If the memory requirements of the system outstrip the memory available on the chip it seems most likely that the wrong device has been chosen. However, there are applications which require extended memory, beyond the capabilities of any device on the market. For example, devices which have to store large amounts of data such as data loggers and personal electronic diaries, may require an external RAM. Data and address buses, as discussed above, are not available externally on a 4 bit microcontroller, and so the ordinary I/O lines have to be used to transfer addresses and data to the external memory device. A 64kbit memory, organised as 65k × 1, uses 20 I/O lines – 16 address lines, data in, data out (one line each), chip enable, and write enable. As the data link is in effect a serial link it imposes a software overhead, and so perhaps a 16 × 4 arrangement would be better. This uses the same number of I/O lines. The address lines reduce by two, but the data lines increase by two, the data lines being bidirectional in the 4 bit wide part. Going up to 8k × 8 might appear to make sense in some applications where data is stored in bytes, but the penalty is the requirement of 4 extra data lines, and only losing one address line. The 16 × 4 part seems well matched to microcontroller. This would probably be a static RAM, especially in battery operated equipment, because of its lower power requirement, and the fact that it is easier to provide battery back-up, needing no refreshing. Dynamic RAM does have the disadvantage of needing to be refreshed, a fact which adds a considerable overhead in the comparatively slow world of 4 bit microcontrollers. The great advantage of dynamic RAMs is the mutliplexed address structure which saves on I/O pins, and allows bigger memories to be accessed.

Apart from the need to access large amounts of RAM, there are some other situations where an external memory of some sort is required. Some system parameters may be variable, but have to be maintained, once set, through on/off cycles, power loss, battery changes, etc. In this situation an EEPROM is very useful. There are some parts available which include on-chip EEPROM, but it is quite likely that a particular application will demand features they do not possess, and so external EEPROM will have to be used instead. Most of these come in byte wide arrangements, but there are some serial parts available which save on I/O lines, but are obviously slower (see Data buses above).

5 Hardware design II – The operator interface, A/D and D/A conversion, and sensors

This chapter considers three aspects of hardware design, the human interface, including displays, keypads and bleepers; the topic of conversion between analogue and digital values; and the subject of sensors. All these aspects of design are important for systems operating in a real world, and it is an unusual system which does not have to touch on any of these. The material in this chapter is complemented by discussion of software aspects in Chapter 8, and there will of necessity be some overlap between the two as hardware and software are so intimately linked.

Displays

From a few lights on an indicator panel to a complex LCD, the display that a microcontroller drives is crucial to the operator. It is usually the principal means whereby the system can give information to the operator. This section deals with LED, LCD, VFD and CRT displays, which are the display types mainly used in microcontroller systems.

LED displays

Light emitting diode displays are used where a high brightness display is required, and they are useful as single indicators, as seven segment displays and as bar graph types. The commonest colour, and the least expensive, is red, but green is also common and blue is also available, so any desired colour can be generated. LEDs require currents in the milliamp range, from about 1 to 20 depending on the efficiency of the LED and the brightness required, and this level of current cannot be handled by the standard CMOS output used on most microcontrollers, but many devices do have high current output ports which can be used to drive LEDs. Alternatively, an external transistor can be used to give the necessary current gain. LEDs also require current limiting resistors, as they are essentially constant voltage devices, usually operating at about 1.5 volts. A common technique useful particularly with multi-segment displays is display multiplexing. Instead of all the diodes being taken to a common cathode connection, they are arranged in groups, or phases, and for each phase there is a switch. An output port can then be connected to the anodes of several diodes, each on a different phase. The phases are switched on in sequence, only one phase being on at a time, and each anode drive output has to be loaded with the information for the 'on' phase at each switchover point. The result is a flickering display, but provided the flicker is of a high enough frequency the eye cannot detect it. The appearance is of a display which is receiving the same average constant DC current. In fact, sometimes a multiplexed display can seem brighter than a continuously driven display receiving the same average current.

LED displays with large numbers of segments are more appropriate for small volume production, and as microcontrollers tend to be aimed at high volume markets, they tend not to provide internal circuitry for multiplexing LED displays. LCD display drivers almost universally provide a multiplex facility, but the requirements of LCD are very different from LED, so it is not appropriate to try to convert LCD signals. It is not too great an overhead to provide the multiplexing circuitry, and it can sometimes be combined with the keyboard drive so that the software can be simplified (see Fig. 8.6).

Liquid crystal displays

The LCD has established itself as the major display technology for low power systems, and as a result nearly all the microcontroller manufacturers have integrated drivers to support LCDs. The LCD consumes practically no power, but requires an alternating voltage drive. LCD displays can be damaged by DC components in the drive voltages, and so LCD drivers are carefully designed to avoid this. The LCD is usually custom designed for the system, but there are a number of standard LCD types available. In the main, these fall into two categories. There are non-multiplexed seven segment displays, providing a number of seven segment characters, probably from 3 to 8, plus a few extra symbols, such as decimal points, colon, and plus and minus signs. LCD drivers in 4 bit microcontrollers can usually be programmed to drive this type of LCD, but they are normally designed to drive multiplexed displays, so it may be difficult to find a microcontroller with enough segment lines to drive a large non-multiplexed display. There are also the heavily multiplexed dot matrix modules which have full alphanumeric capability, and can usually operate in graphics modes. Not many 4 bit microcontrollers are capable of driving the number of segments to be found in one of these dot matrix units, and drivers are not usually capable of working to the level of multiplexing required. Instead, the units are supplied complete with driver chips, and sometimes a character generator, and communication with the module is via an 8 bit data bus, making them less suited to the 4 bit microcontroller world.

If it is necessary to use a custom made LCD, there are a number of points to consider when drawing up the specification. The multiplexing feature of an LCD is built in at the manufacturing stage. An LCD display consists of a thin layer of the liquid crystal material sandwiched between two sheets of glass. On the glass are transparent electrode patterns, and when the appropriate alternating voltage is applied between electrodes on opposite faces of the device, the liquid in between is organized to rotate the plane of polarisation of the light passing through it. A polarising filter on the front of the device, and a reflector on the back, complete the display so that light is only obstructed where the drive voltage appears on opposite electrode areas. The pattern on the back piece of glass is referred to as the backplane, and this will be divided into as many parts as are required by the multiplexing scheme. For most 4 bit microcontroller LCD drivers the limit is four sections of backplane, and it is usually possible to configure the driver in software to drive non-multiplexed, biplexed, triplexed and quadriplexed displays.

LCDs are very sensitive to applied voltage, and when drawing up the specification of a display the operating voltage is an important parameter. Different liquid crystal materials, and different electrode spacing are employed to ensure that the display gives the necessary density at the design voltage. The displays are also sensitive to viewing angle, and so this has to be specified at the outset of the design. In order not to apply any DC component to the display, and still be able to multiplex it, drivers make use of the fact that a small applied alternating voltage will produce no visible effect, and that the display switches over a relatively narrow voltage range. Figure 5.1 shows a typical set of drive signals for a quadriplexed display. This is a one third bias voltage scheme, where the off segments are subjected to one third of the voltage of the on segments. It is up to the manufacturer of the LCD to ensure that the LCD shows nothing at the one third

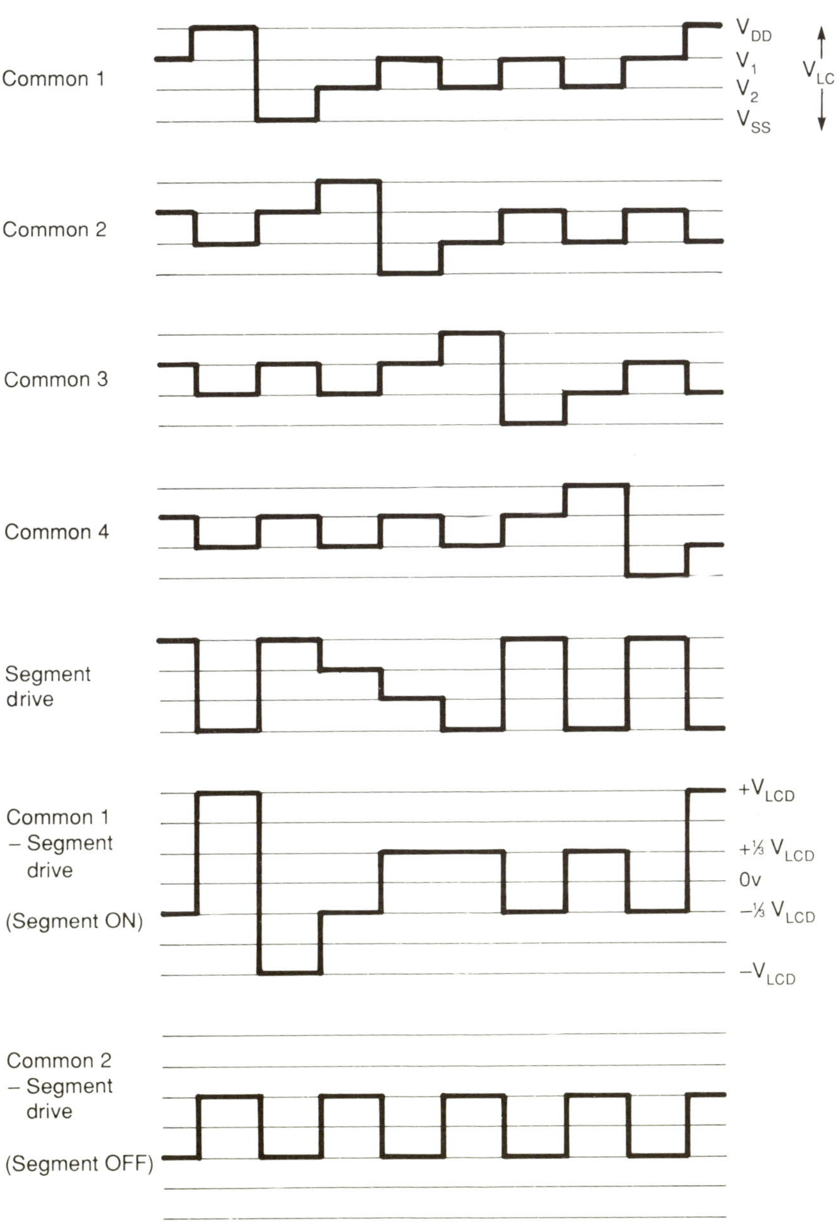

5.1 Typical LCD drive waveforms.

voltage drive, but will display good, dense segments at the full drive voltage. One other parameter required by the LCD manufacturer is the frame rate, the number of frames, or complete cycles through the phases, that are completed per second. This will be dictated by the LCD driver in the microcontroller, but there may be several options according to how the microcontroller is being used.

Vacuum fluorescent displays

This type of display is widely used on domestic equipment such as video recorders and clocks, and also in automobiles. They give a clear, easy to read, self-illuminating display, but have the disadvantage of being comparatively power hungry. Where they are used on some calculators it is usually the mains powered desk models rather than the pocket, battery operated types which use VFDs. The VFD operates by having a high voltage (25 to 40 volts typically) between the filament and the anode. A grid is used as a control element to allow multiplexing and to control brightness. The high voltage used by VFDs means that they can only be used with microcontroller ports which have been specifically designed for the purpose. VFDs are available in a wide variety of configurations, and there are many microcontrollers with integrated VFD drivers, which will take care of the multiplexing, brightness control, etc. In some parts special provision is made for scanning the keyboard while driving the display, as this is a useful way of organizing a system which has both keyboard and display. If it is necessary to use a VFD with a microcontroller not designed for such use, there are separate display controller chips available, but to have to resort to their use rather defeats the object of using a 4 bit microcontroller.

CRT and TV displays

The CRT (cathode ray tube) is used universally as the display medium for computers of all sorts, but the types of application where 4 bit microcontrollers are used tend not to require the kind of display capabilities of the CRT. However, there is a class of applications where display on a CRT is required, and that is in connection with the television set. There are a number of 4 bit microcontrollers available which have the facility to allow numbers or text messages to be displayed on a TV screen, probably superimposed over the existing picture. This is a rather specialised application used mainly in TV tuners, satellite decoders and VCRs.

Keyboards and keypads

The commonest way for the operator to communicate with a system is to provide some buttons for the operator to press. These may be special function buttons, such as channel select on a TV tuner, where a button is dedicated to a unique function (e.g. select channel 1), or they may be more general purpose such as on a numeric keypad, or a 'qwerty' alphanumeric keyboard. Sometimes special function keys will change their function according to the circumstances. On a watch, for example, where the number of keys is limited, one key may be used to select a mode of operation, and the function of other keys will change with the mode. A key which in one mode is used to change the time, might in another mode be used to stop or start a stop-watch. Whatever their application, there are some common features worth noting. First, the action is usually momentary, not latching. The keypress has to be detected while the finger is on the button. Also, the operator expects an instantaneous response. Instantaneous in this situation probably means within a few tens of milliseconds. Response time is primarily a software concern, but hardware arrangements obviously have some bearing on this, perhaps affecting the choice of I/O port to use for the keyboard interface.

The types of switch attached to the key also have some bearing on the way it is used. Very common in many applications is the conductive rubber type, where a low resistance path between tracks on a PCB is created by pressing a conductive rubber pad onto the PCB. The resistance of this type of switch is usually a few hundred ohms, and designs should allow for 1k ohm. A pull up or pull down resistor on the input line ensures that it takes the correct value when the switch is open, and when the switch is closed the resistance of the switch comes in to pull the input the other way. If the pull up/down is not of high enough value there could be a problem. Where a microcontroller provides internal pull up or down resistors their value is sometimes rather low, and if the switch resistance is high, this introduces problems of marginal input voltage values, and poor noise immunity. Conductive elastomer switches are comparatively good in terms of 'bounce', but there are some metal contact switches where this can be a problem, and careful design, both in hardware and software, is required to ensure correct operation from the switches.

If there are only a small number of keys in the system, eight or less, it is usual to allocate one unique input line to each key. The keyswitch can either switch the input to the positive or the negative voltage rail. With CMOS inputs it makes no difference to the current. There will also have to be a pull up or pull down resistor on the input to ensure that the input sits at the correct polarity when the key is not pressed. Sometimes there are internal pull up/down resistors available on the microcontroller chip, which is useful in reducing the external component count provided the internal resistor value is suitable for the application. The internal pull up/down resistor is usually either a mask option or a program option. In some situations it is better to use external resistors in order to get a value which suits the application better. This is particularly true in battery operated, low power products, where too low a value of resistor can lead to excessive current drain while the button is pressed leading to undesirable effects such as battery voltage depression and display fading. Capacitors can be connected in parallel with the switches if switch bounce is a problem, but care will be needed in selection of capacitor and pull up/down resistor values to give the correct time constant.

If more than eight keys are needed in the system then a matrix approach is usually employed. The way this operates, together with the software considerations of such an approach are examined in Chapter 8. Also covered in Chapter 8 is the technique of combining key scanning with display driving to make the software as efficient as possible.

Bleepers

Systems often require some audible output, and many 4 bit microcontrollers cater for this by providing oscillators or melody generators on board. For low power systems the challenge is always to provide sufficiently loud output without excessive power consumption, and the standard technique is to use a ceramic resonator or piezo electric device. Such a device is mainly capacitive in its loading, and consequently requires fairly low currents. Even so it is often necessary to provide an external gain stage to boost the output capabilities of the microcontroller. Piezo electric transducers usually exhibit strong resonance, and to get the most out of one it should be driven as close to resonance as possible. Mounting the transducer, or enclosing it in some way can alter its frequency characteristic, reducing the 'Q', or even moving the resonance away from that of the unloaded condition. Including some inductance in the gain stage helps to boost the output by elevating the drive voltage.

If bleeps of more than one frequency are required, and particularly if melodies have to be generated, using the transducer at resonance will not be possible, and having the resonant frequency in or near the range of the notes of the melody could cause peculiar effects, so the transducer has to be chosen and mounted with care to ensure that the desired effects are produced. Piezo electric transducers are usually small devices, from about 10 mm to 30 mm in diameter, and so are more effective at the upper end of the audio frequency spectrum. A common frequency for single frequency resonant operation is 4 kHz. Systems with more power available can operate larger moving coil speakers which give better reproduction at the lower end of the spectrum than piezo electric devices, and have less pronounced resonance effects. They are more suited to applications such as speech synthesis, where a wide range of frequencies has to be covered.

Analogue to digital conversion

Most manufacturers have parts in their product ranges which perform A/D conversion in one way or another. The usual method is by successive approximation, in which the input of a D/A converter is set bit by bit, and the output compared at each stage with the analogue input to be measured. There are some other types available, but successive approximation is by far the commonest. Some have a similar scheme, but using a counter input to the D/A which is inherently slower than the bit by bit method. Another method, for measuring resistance, which is particularly applicable to temperature measurement, is to use the resistor to control the frequency of an oscillator, which is then measured using the chip's counter/timer facilities. An allied method, using time measurement, is the dual slope method, where the unknown voltage is integrated across a capacitor for a fixed time, and then a known voltage is used to integrate in the opposite direction until the starting voltage is reached. The ratio of the two times then gives the unknown voltage. This method is used on at least one device on the market.

Some parts just provide analogue comparators for use in situations where a full A/D conversion is not required. Most parts integrate the control logic so that all the program has to do is set the conversion going and collect the result when it is finished. Texas Instruments, though, in their TSS400 family, provide all the tools but leave it to the software to perform whatever conversion algorithm is required. National Semiconductor do not have any integrated A/D conversion circuits integrated into their COPS family, but their data book has a good application note on building converters around their parts, which compares the various techniques, and gives an indication of the speed and accuracy that can be achieved with each. Speed and accuracy are key parameters when choosing an A/D circuit for use. A successive approximation converter will clearly take longer if it has more bits to work through, and so a fast, accurate conversion will only be available on higher end microcontrollers with faster clock rates. Faster usually means more power, and so choosing the right part becomes a balancing act, trying to strike the best compromise between the various parameters.

Power supplies and reference voltages are very important when seeking to perform accurate analogue to digital conversions. Basically, an A/D converter is calculating a ratio between an input voltage and a known reference voltage, and expressing it as a binary number. Unless the reference voltage really is known, the results of the conversion will be of little use. When it comes to accuracy, the conversion can be no more accurate than the reference voltage. Many A/D converters are susceptible to errors caused by digital noise appearing on the reference voltage, or on the voltage being measured. Some parts use the same Vcc supply for the A/D converter as for the rest of the chip, and also use it as a reference voltage, so that any measured voltage is expressed as a function of the supply. If the supply is noisy this can give rise to poor quality conversions, and so great care must be taken with decoupling and track routing to ensure minimum interference, and also minimum impedance to noise signals returning to ground. Other devices give alternative power input pins for the A/D converter section of the chip, which gives more scope for isolating the analogue supplies from the noisy digital supply. The reference voltage is sometimes taken to be the supply voltage, and sometimes a separate input is provided. Where there is a separate input it does allow the user to make use of a stable reference voltage source, but care should be taken when making such provision to ensure that the reference supply can operate correctly into the given A/D. The reference voltage is usually applied to a resistor ladder, from which taps are selectively switched to the comparator input. This tap switching is a source of noise if the reference voltage has too high an impedance. Note that the result of the conversion, the binary number in the A/D converter's output buffer, is a function of the reference voltage, and so some further arithmetic may be required to turn the answer into a number of volts, millivolts, or whatever may be required.

Another possible source of trouble in A/D converters is crosstalk between analogue inputs, where a circuit has a multiplexed analogue input, as most do. It should not be too great a problem if care is taken with track layout and ground return paths. One problem to which some devices are susceptible is out of range inputs causing other inputs on the multiplexer to vary. The data books will give absolute maxima and minima beyond which inputs must not be taken, as permanent damage to the device may result, but it may be found that within this permissible region, especially if the input voltage goes outside the supply voltage range, there may be input voltage ranges which should be avoided to ensure no interference from one input to another. It may be sufficient to include catching diodes connected to the supply lines, but there is still the possibility that the diode forward voltage drop will be enough to cause a problem, and other means may have to be found to ensure that the input voltages cannot go out of range.

It is usually important that an input voltage does not change during the conversion. In a successive approximation converter, the most significant bit is determined first, and there is a comparison with the input voltage for each bit of the result. As time is required to work through the bits, there is the possibility that the voltage being measured at the end of the conversion, when the least significant bit is being found, may be a different value from the voltage that was being measured at the start of the conversion. To overcome this problem a sample and hold circuit must be used if the signal bandwidth and the conversion time combine to create a situation where the input can change ± half the least significant bit during the conversion time. For example, a circuit performing an 8 bit conversion in 40 microseconds can, without a sample and hold circuit, tolerate an input bandwidth of about 15 Hz, so a sample and hold circuit will normally be required. Some devices have them built in to the analogue input circuitry, but they are in the minority. An external sample and hold circuit must usually be provided.

Digital to analogue conversion

The D/A conversion technique almost universally applied in 4 bit microcontrollers is the pulse width modulation technique. In this method a digital output is switched high for a time, t1, and then low for a time, t2. In order to produce a DC voltage, the resulting waveform has then to be put through a low pass filter, and the output voltage is then proportional to t1/(t1+t2). The cycle time, t1 + t2, is usually kept constant, and is determined by the microcontroller in use, and the operating clock frequency. There may be some choice of values, but the highest possible frequency is usually required to avoid large component values and long time constants in the low pass filter. For a 14 bit PWM, the basic frequency is usually in the range 120 to 150 Hz, but several manufacturers employ techniques to effectively increase this. A typical method is to divide the fundamental frame into 64 subframes. A pulse is output in each of these subframes, whose width is determined by the value in the top 8 bits of the 14, and the lower 6 bits determine how many of these 64 pulses should be extended by one basic interval of time so that the average DC value of the waveform is correct. This gives an output waveform with a fundamental frequency in the 8 kHz region. Clearly this method allows the low pass filter to use much smaller components, and gives the system a much quicker response time than would be possible with a fundamental frequency of 120 Hz. The actual design of the low pass filter is beyond the scope of this book, and is highly application dependent. It may be that a simple series resistor and parallel capacitor can provide the necessary smoothing, or for more demanding applications a design with a sharper cut off may be required.

In most applications the output from the microcontroller will drive a switch of some sort, probably a bipolar or field effect transistor, as the current drive capability of the PWM output will probably be insufficient for the application, and the logic levels not suitable to be used as references for the control of the voltage to be generated. Instead, a stable reference voltage, often higher than the Vcc of the microcontroller, will feed into the switch which is controlled by the PWM output, the output of the switch feeding into the low pass filter.

In many systems the PWM output is used to control a motor drive, and increasingly that motor is a brushless DC motor, with its inherent advantages of high acceleration and freedom from brush noise. In order to drive a brushless DC motor, the supply has to be commutated outside the motor, as there are no brushes/commutator rings to do the job. There are now dedicated driver integrated circuits available for driving such motors, and the choice has to be whether to provide the commutation logic in software inside the microcontroller, or whether to leave it to a dedicated driver chip. In view of the fact that such a chip will often be able to accept an unsmoothed PWM signal directly as a speed demand input, and will also have quite powerful output drive capabilities, the dedicated driver may well be the most cost effective solution.

In some situations the PWM facilities provided by a microcontroller may not be suitable for the task to be performed. For example, in controlling the brightness of a lamp which is supplied with power from the mains electricity, the drive from the microcontroller to the thyristor is a form of PWM, but it has to be synchronized with the mains electricity frequency in order to operate correctly. Here it will be necessary to use software routines to perform the PWM operation, perhaps using a comparator input on the microcontroller to give zero crossing detection.

There are some other methods of digital to analogue conversion available in 4 bit microcontrollers. For instance in the NEC μPD17102 there is a 6 bit converter which works on the resistor chain technique. A chain of 64 equal resistors is connected between the analogue ground and the reference voltage, and a 64 way switch selects the tap on the chain corresponding to the value to be converted, and directs that voltage to the output. This gives a much quicker conversion, but its use is limited to lower resolution converters because of the increasing complexity of the method when applied to longer digital words.

Position sensors

Many systems in which 4 bit microcontrollers are used employ position sensors of one sort or another. A daisy wheel printer, for example, will usually use stepper motors to drive the daisy wheel round, and to move the printhead from side to side, but when the printer is switched on, or receives an 'initialise' command, it has to find its home positions, and ensure that the daisy wheel is properly engaged with its drive mechanism. To do this there will be sensors which indicate the home positions. Position sensors come in a variety of types. Optical sensors are very versatile, and are commonly used on office equipment, in industrial control, and in a wide variety of other application areas. There are basically two types. In one type the light source (usually an infra-red LED) shines directly at the sensor, and in the other type a reflector is required to direct the light on to the sensor. In the first type, the desired position is found by a blade of some sort entering the slot between light source and sensor, hence cutting the beam of light and allowing the output from the sensor to change. With the reflective type the same effect can be used, having a fixed reflector, and detecting when the beam is cut. Often, though, the reflector is mounted on the moving part, and then the presence of the signal from the sensor indicates the required position.

Driving the sensor and the light source is usually very simple. The source can often be permanently powered from the supply to the microcontroller, but sometimes it is useful to have a switched supply, controlled by the microcontroller. This gives more scope for diagnostic programs to detect failure situations. It may be necessary to provide the light source with more current than the microcontroller can provide, but a single transistor gain stage usually suffices. For the light sensor, all that is needed in the way of drive circuitry is a pull up or pull down resistor in many cases. More sophistication is sometimes required, particularly if ambient light is likely to cause problems which cannot be overcome by suitable light filtering on the window of the sensor. It may also be necessary to be able to accommodate a range of levels of light from the source. In this case there may have to be some sort of automatic gain control built in to the sensor drive. The microcontroller itself might be built in to the gain control loop, another situation where it is useful to be able to switch off the light source. The light source is switched off to allow only ambient light on to the sensor. It is then switched on again and the sensor response checked, and gain adjusted if necessary. Another technique sometimes employed if interference from ambient light is a problem is to modulate the emitter with a known signal, and look for a response from the receiver which is in phase with the modulation. This is useful if the ambient light is not constant, a situation which is quite common as many sources of ambient light flicker at the mains frequency. One of the advantages of the optical sensor is its speed of response, and care has to be taken with the sensor drive that suitable time constants are chosen to suit the application.

Inductive proximity sensors are another important group of sensors which are likely to be found in 4 bit microcontroller systems. The sensor contains a coil whose inductance will change if the sensor is brought close to a metal object. It also usually contains the necessary drive electronics to give a digital output indicating the presence or absence of the target within the sensing range of the detector. Sensors are usually available with either PNP or NPN output transistors, so a system designer has the choice of polarity. The only external components then required are a pull up or pull down resistor, and a line to the supply voltage. Some units will operate with a supply as low as 5 volts, but many require a higher voltage drive, introducing the minor problems of interfacing to the microcontroller – usually only a matter of a couple of resistors and a catching diode for safety. The inductive sensor operates using an oscillator whose frequency is determined by the inductive coil. Any change in inductance, caused by the proximity of a metal object, will cause the oscillator frequency to change. Usually the oscillator, amplifier and frequency threshold sensor circuits are contained in the device, but for some purposes, for example with intrinsically safe equipment for use in hazardous areas, the sensor contains only the oscillator components, and external circuits are required for amplification and frequency thresholding.

Allied to the inductive sensor is the capacitive sensor which has the advantage of being able to sense such materials as wood, glass and PVC. The capacitive sensors are usually completely self-contained, requiring only a supply voltage, and giving a binary output. They tend to be more expensive than inductive sensors, and there is not such a wide choice of types available, but interfacing them to a microcontroller is similar.

There is a variety of other types of sensor available, giving basically a binary output, and having differing properties, making them suitable for different applications. Ultrasonic sensors and Hall effect switches are examples. In each case an output is available indicating the presence or absence of the target, and this gives a binary input to an input port on the microcontroller, allowing the control program to take appropriate action. Such sensors are often coupled to interrupt inputs, allowing the section of the program concerned with the input to be obeyed as soon as possible after the triggering event.

Apart from sensing the proximity of a target of some sort, which is frequently used in all sorts of motion control applications, there are other binary type sensors which can be used to sense other physical phenomena, and are interfaced to the microcontroller in a similar way. The sensor appears to the system as a switch which can be open or closed, and the pull up or pull down resistor determines which internal representation, 1 or 0, represents which state of the switch. Microswitches, magnetic reed switches, mercury tilt switches and light operated switches all come into this category. The controlling software may have to make allowance for the dynamic behaviour of the switch, by 'debouncing' the input, or it may be necessary to include some capacitance in the circuit to reduce noise pick-up from a long line connected to an open switch.

Multiple value sensors

The sensors discussed above are binary in nature. They have two senses, on and off, and consequently their interface to a digital device such as a microcontroller is a relatively simple affair. In many systems, though, there is a wide variety of analogue quantities that a microcontroller may be required to measure. The discussion on analogue to digital converters above gives some feel for what is available from a 4 bit microcontroller in terms of its ability to handle analogue inputs. The tables in Chapter 3 show that there are many devices with an 8 bit conversion capability, usually multiplexed between a number of analogue inputs. Each type of sensor, be it temperature, humidity, pressure, strain gauge, load cell, etc, requires its own type of signal conditioning circuit. This converts the primary variable quantity, such as resistance or voltage, into a voltage in a range suitable for input to the microcontroller analogue input. The signal conditioning circuit often also provides a linearising function prior to conversion of the analogue quantity to a digital value in the microcontroller, but this function can often be performed by the microcontroller software to be more cost effective. Signal conditioning circuits for the common types of sensors are available as integrated circuits, or perhaps hybrid modules, which are easily assimilated into the microcontroller system. Care has to be taken, of course, as with any mixed digital/analogue system, to ensure that digital noise does not interfere with the analogue signals. A separate analogue ground return line, connected to the digital ground at only one place, is a usual precaution.

There is another class of sensors which give multivalued outputs, but present these in a digital form. For example encoders indicating shaft angle operate in this way. The interface to the microcontroller consists of the appropriate number of lines into input ports, perhaps with pull up or pull down resistors. The code presented by the device may not be the straightforward binary value, as it may be encoded in some way. For example, Gray code is often used in absolute encoders because it has the property that only one bit on the interface changes at a time as the angle changes by one unit of resolution. This has the advantage of minimising errors introduced by skew. At the point where the encoder changes from one code to the next, it is possible that some bits may change fractionally ahead of others, because of slight misalignments of the reading head, or errors in the encoding disk. If the microcontroller samples the encoder output just as it is changing from one segment to another, it may pick up some of the bits from one segment and some of the bits from the next. If only one bit changes there is no serious error, but if more than one bit were to change, then there is the possibility of gross error. Gray code will have to be converted to binary by the software before calculations can be performed, or, if a look up table approach is being used, rather than calculation, the input code can be used directly to address the look up table.

Operator inputs often appear as multi-bit digital signals. Set points and other control variables are often input via thumbwheel switches, or similar devices. Again, a simple arrangement of pull up or pull down resistors provides the necessary interface. Often the input will have to be converted from BCD before use, and it may be necessary to take special precautions when numbers are being changed, as the operator action will be very slow compared with the microcontroller reaction, and there will probably be a number of intermediate values appearing on the input before the final desired value is reached, and acting on the basis of the intermediate values might cause a problem. The simplest arrangement is for the operator to press a key to indicate that a new value should be read, but other schemes, friendlier towards the operator could be devised, such as acting on a new value after a predetermined delay.

Another class of input devices becoming commoner are code readers of various types. Cards with magnetically stored information, and printed bar codes are examples of the types of code that might be encountered. Readers are available to system designers which include a degree of intelligence, possibly provided by a 4 bit microcontroller. Such devices connect into a system using a standard interface, such as RS232 serial communications. However it may be desirable to work at a closer level, working directly with the data from the magnetic or optical pickup. After some suitable amplification this becomes a binary input, and the software then has to take care of interpreting the data, timing, decoding, applying error checking, etc, and finally informing the operator if the reading operation was successful.

System outputs

Much of the ground concerned with system outputs has already been covered under other headings. Often the main system output is the display, and the common types of display have been discussed. There are other types of display, for instance electromechanical displays, which could be regarded as the type of output which is discussed in this section. The section in Chapter 4 covering communications and data buses also covered a large number of possible output types, and the paragraph on digital to analogue conversion earlier in this chapter covered a bit more of the ground. However, there are a number of types of output which have not been discussed at all, and are worthy of mention.

Many system outputs are single binary outputs, operating switches of some sort, and where the output is a digital numeric output it can often be regarded as a group of single binary outputs. In most cases the microcontroller is buffered from the eventual load device by a switch of some sort, be it a solid state device, or an electromechanical switch. Driving this buffer often requires some current gain, usually supplied by a transistor. In some systems the final load device is quite close, in circuit terms, to the microcontroller, requiring just that single current gain stage as a buffer. For example, the print head on a dot matrix printer, where each wire of the head is driven by its own electromagnetic coil, could be driven by a microcontroller using a high gain darlington transistor. One characteristic common to many of these loads is that they are inductive in nature, and as such require careful handling. The main problem with inductive loads is switching off the current, as the energy stored in the magnetic field requires the current to continue to flow in order to allow it to dissipate. How much of a problem this is depends on the amount of energy stored, and how quickly it has to be dispersed, usually to be radiated away as heat. If the inductive energy is very small it can sometimes be absorbed by the switching transistor. The back emf generated by quenching the current causes the transistor to go into some form of breakdown from which it can recover. Care has to be taken to ensure that the transistor can absorb this energy, and that the heat generated can be dissipated by the physical arrangements provided. More often though, it is necessary to provide some sort of flywheel circuit, with a resistor to absorb the energy, and a diode to prevent the current flowing through the resistor except when the back emf is being generated by the inductive load. Precautions here must ensure that the back emf is not going to rise to a level which would cause damage to the transistor. One problem with this arrangement is that the current can take a long time to decay, and the mechanical action being controlled can be adversely affected by the residual magnetism.

Sometimes it is important to quench the magnetic field in as short a time as possible. The dot matrix printer head is a case in point. Sufficient current must be given to drive the wire forward against the spring pressure to ensure that it hits the ribbon hard enough to print a good dot, but it must withdraw quickly as the head is on the move and a delay can cause smudging. One possible solution to this problem is to replace the resistor in the flywheel circuit by a zener diode. This has the effect of maintaining a nearly constant back emf for as long as the current continues to flow after the moment of switch off, and ensures that the magnetic field is quenched as fast as possible. The inductive energy is mainly dissipated in the zener diode, so it must be of sufficient power handling capability to cope with this. The value of the zener diode can be chosen to match the breakdown voltage of the switching transistor, and so making sure that the transistor is protected against this type of failure.

6 Software design I – An introduction to software and some peculiarities of 4 bit microcontrollers

The software design aspects of 4 bit microcontroller systems are of major importance when seen in the light of overall system development, so a number of chapters have been devoted to this aspect of system design. In the text of this and succeeding chapters numerous examples of microcontroller codes will be found. The aim of these coded examples is not so much to present the reader with ready made solutions to drop into systems, but rather to present ideas, techniques which can be employed, methods of approaching problems which may arise in designing a system. There are a great many different types of 4 bit microcontroller, and although they are often grouped in families which share identical, or very nearly identical instruction sets, there are still a great number of different device architectures and instructions sets around. Because of this, no attempt has been made to cover the field, but it is hoped that the variety of instruction sets used in the examples will be sufficient to show techniques to use in a big enough variety of types to be useful. It is left to the reader to translate the examples into something exactly right for a particular microcontroller in a particular system. Often the examples are incomplete, showing just sufficient code to illustrate the point under discussion.

While an attempt has been made to give examples in code from a wide variety of types of microcontroller, the astute reader will soon see which types of microcontroller the author is particularly familiar with. The bias that may appear in the coded examples towards one manufacturer over another is not intended as a value judgement, it is merely a matter of familiarity, and of being able to present examples which have, on the whole, been used successfully in systems. Not all the examples are from 'real live' code, though; some are generated purely by reference to manufacturers' data books. Little use has been made of flow charts, as it is felt that the coded example, adequately commented to ensure that the peculiarities of an unfamiliar instruction set are not a barrier to understanding, will in fact convey more accurately exactly what is required from a particular routine or section of code. The reader may find it useful to generate flow charts for the coded examples as an aid to understanding and translating into the instruction set of another microcontroller. It should be said here that the author makes no claim to perfection, and hopes the reader will enjoy improving on the coded examples given in the text.

Software development

When a system controlled by a 4 bit microcontroller is being developed there are several aspects which have to be considered, but it is likely that developing the software will be the biggest single aspect, and the one which takes up the most time and effort. In the light of this, some understanding of the software development procedures will be useful in planning such a development.

Figure 6.1 shows a typical product development process. Software and hardware design are started in parallel, and there has to be interaction between them from the start. Design is often carried out in sections which can be tested and proved on their own before they are added to the final system. This is particularly true of software design, which is usually developed on a modular basis. There is sometimes a requirement for special software to assist in hardware development – special routines to exercise the hardware in a particular way – but this is separate from the mainstream software development, although test routines may well be a part of the final software requirement. There will also be a need to provide the software designer with some sort of test bed on which to prove out the software. The whole process is iterative, going back to redesign parts which do not perform to specification, and sometimes modifying the specification as experience is gained. It may be that some detail in the specification has not been thought through properly, or it is inconsistent in some respect, which only shows up as the product or system begins to take shape. It may even be that the specification was too ambitious in terms of what could be achieved with the available equipment, or in a given budget. It is important to be aware at the outset of any development that the initial specification is a statement of what is desired rather than a hard and fast statement of what is to be achieved. The 'customer', whether that means senior management of a development group, another department of the same company, or an external customer, must not be allowed to embark on a development project without being aware from the outset that it may not be possible, or even desir-

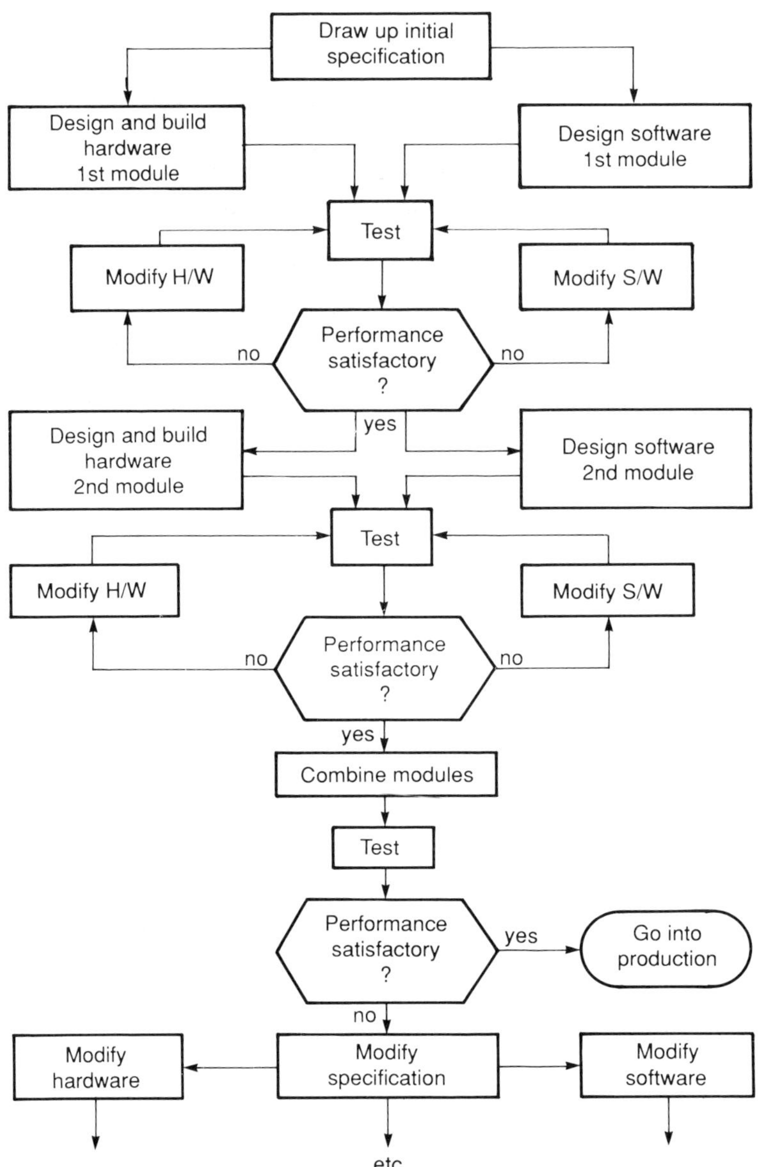

6.1 The software development process.

able, to achieve everything that is originally intended. Any development plan must allow for changes to the specification during this iterative process of designing and testing the product or system.

When it comes to the software, the approach to designing this is usually modular. In systems controlled by the more powerful processors, the software development environment is geared up for this, and all sorts of software development aids are available to assist with controlling the development in a modular fashion. However, 4 bit microcontrollers are small, simple devices when compared with other microcontrollers and microprocessors, and because of this they place various restrictions on the software used in them, and the way it is developed. Modular programs are often not directly supported. Programs are written in assembly language, and the assembler is absolute, and there is no link facility. Many of the advances in design seen in the more powerful 8 and 16 bit devices have been specifically aimed at making it easier to develop programs for them – instructions sets designed to suit high level language compilers, the provision for cache memories, and other programmer friendly features, which have trickled down from large mainframe number crunching computers into the world of microcomputing, have not yet reached the realms of the 4 bit microcontroller, nor are they likely to, except at the very top end of performance (and hence price). The primary raison d'etre of the 4 bit device is low cost, and to keep the cost to a minimum it is necessary to minimise various aspects of the design, and hence place restrictions on the user. This may make system development a little more tedious, as the programmer does not have available some of the time saving features to be found in bigger systems.

In the final system in which the software under development is to operate, the 4 bit microcontroller will probably be mask programmed if the system is to be produced in a reasonable volume. The program is 'built in' to the microcontroller during the manufacture, and is fixed. It is important therefore to get it right, as changes involve remasking which is a costly procedure. The problem of getting software right is not unique to 4 bit microcontrollers, and the general methodologies which have been devised to help towards writing correct code can often be directly applicable to the 4 bit microcontroller situation. A well planned, clearly structured program will be much easier to test and correct, and it will be much easier to generate tests for a well thought out specification. Testing the system must be part and parcel of the development plan, and plenty of time must be allowed for it – it usually takes twice as long as expected. It is worth noting here that many development environments do not model the final system exactly, especially if there is no EPROM, or other field programmable version of the chosen microcontroller. Consequently there are aspects of the system design which cannot be tested fully before masked parts are made. If field programmable parts are available, the preproduction models, and perhaps the early production versions, can be manufactured and tested in their final form before committing to the mask. In this case there is a good chance of getting it right first time. If this is not possible because of the lack of a field programmable version it would be wise to budget for two attempts at masking. It is not until the customer has the final product, or something very like it, in his hand, that he really discovers what he has specified. So even if the programming is quite correct according to the specification, there may well be changes at a later stage. If the price the customer is expecting to pay has already made allowance for a second mask, it will make it much easier for everyone involved to get the final product right.

One of the first things the programmer new to 4 bit microcontrollers has to get used to is working with 4 bit data words. It may also seem strange that the instruction word length is not restricted to 4 bits, and can in some devices be up to 16 bits. This difference can take some getting use to, particularly when dealing with addresses, both program and data addresses, as sometimes they are embedded into instructions, and at others they are held as data. This gives rise to a strange mixture of addressing ranges, some instructions addressing the complete code space, others perhaps only half of it. The programming manual has to be studied carefully to make sure that all the restrictions this causes are fully understood.

Assemblers

The 4 bit microcontroller is essentially an assembly language device. One or two manufacturers support compilers of some sort, and some of the parts available at the top end of the range have aspects which permit the use of C (which can be regarded as a mid-level language), or some other suitable higher level language, but the vast majority of systems are developed in assembler. In order to use assembler, the programmer has to be familiar with the instruction set, and also the architecture, of the device in use. Every device is different, so working with a number of different devices can be tiresome, and sometimes confusing. Most of the manufacturers produce parts in families – a single core processor with different amounts of memory, and different integrated peripherals, but a common instruction set. This obviously makes life easier, both for the user and the manufacturer, and encourages users to stick to the products of one manufacturer. Having become familiar with the architecture and instruction set of one device in a family there is a much shorter learning curve to get to grips with another member of the same family, than to change to a part made by a different manufacturer, and using a completely different structure. However, the problems must not be overstated, and choosing the right part must not be unduly influenced by the fear of having to go up that learning curve for a new family of devices.

To encourage users to remain loyal to their products, many manufacturers build attractive features into their assemblers to make them good to use. The provision for the use of macros is fairly standard. A macro is a shorthand way to enter commonly used groups of instructions, with some variable parameters. For example, subroutine calls are often preceded by loading parameters into registers, and the code to display a character 'n' at some location 'm' say, could be contracted from

```
LAI      n      ;load accumulator with value to display
LHLI     m      ;load HL register with location
CALL     DISP   ;call display subroutine
```

to

```
DISPLAY  n,m
```

where the macro 'DISPLAY' has been previously declared as containing the instructions shown in the longhand version. This makes writing programs quicker, and makes them easier to understand. The system has been extended by some manufacturers who provide a number of 'ready made' macros which allow abbreviated instructions to be used. Toshiba call these the 'extended instruction set', and use them to provide a wide variety of conditional branches, subroutine calls and returns, and also multi-bit shifts. The macro definitions themselves often allow conditional statements, so that the same source code assembles differently under different conditions. This is particularly useful if a library of routines has been developed for use in several members of a family, where there may be slight differences in memory addressing ranges, the scope of branch instructions, etc.

Another useful feature to look out for is the provision of libraries of ready made routines to perform some of the commonly required functions on the microcontroller of your choice. Several manufacturers do provide these, and even if they do not provide the exact functions required, they can often be modified quite easily to do so, saving valuable programmer time. They also give a useful set of examples of how to use a particular family of devices to someone new to them and looking for a quicker route up the learning curve.

Assemblers for 4 bit microcontrollers are usually absolute assemblers, as opposed to the relocatable assemblers used with more powerful devices. The programmer decides where the code is going to reside in the program memory space, and the program is put together as a single source module which is assembled to give an absolute object module, and no linker is required. This can be awkward when using a device with a large program memory, and so some manufacturers cater for a certain amount of modularity to be handled by the assembler. On the whole though the programmer is trying to squeeze as many functions as possible into the available space, and needs to know in detail how the program is arranged in the memory in order to use the space in the most efficient way. Sometimes there will be restrictions on where certain items can be located. For instance there may be special areas allocated for use by look up tables, or there may be restrictions on addresses of subroutines making it necessary to ensure that they reside in a specified part of memory. This means that the programmer is usually much more aware of where each part of the code is located than is the case with more powerful microprocessors.

Compilers

Where manufacturers provide compilers, the language used is unique to that manufacturer's products and requires an intimate knowledge of the target microcontroller. The advantage is that programs can be structured in a much more formal way, making it easier to understand what they are doing, and hence making program maintenance much simpler. They also support such statements as IF, THEN, ELSE and WHILE making programs easier to write. One point to note is that when it comes to testing the code, and running the program on an emulator, the diagnostic aids available are not always able to work in the higher language, and to see exactly what is going on the programmer may be forced to work with a listing of the program in assembler. It is worth asking about this before committing to a particular part, and if debugging has to be carried out in assembly language, does the compiler produce an assembly listing? There is more on program development systems in Chapter 10.

Memory addressing

There are various other aspects of the architecture and instructions sets of 4 bit microcontrollers which mark them out from more powerful devices. The simpler architecture usually means that there are fewer data registers available than in bigger machines, and because data is handled in 4 bit nibbles, addressing memory often requires some ingenuity. In some parts it is quite straightforward - the data store is restricted to 256 nibbles, all that can be reached by an eight bit address, and two data registers are used as a pair to provide the address. Often there is also direct memory addressing, part of the instruction, often one byte, is the address field. These methods will be familiar to anyone who has worked with the common eight bit microprocessors, except that the addressing range, and hence the space available in data memory, is much more limited. Sometimes the memory is divided into 'banks' and 'pages', which allows for a bigger data store, but less flexible working. Dedicated registers hold the bank and page numbers currently in use, and a separate instruction is required to set values into each. Addressing within the current bank and page is sometimes restricted to direct addressing. The data location is set in the accessing instruction, and cannot be modified during program running. This is particularly inflexible and makes it difficult to write efficient routines for handling groups of data nibbles. Some parts combine the two methods, using a simple register-indirect method of addressing part of the memory intended for use as general purpose working memory, and a less flexible page and bank system to access a bigger area of memory for less often accessed data. For example, a part intended for use in an extended feature telephone system may have a large area of RAM to store all the data required for a repertory dialling system, but this is infrequently accessed and can afford to use a slower, less efficient addressing system, while there is still a need for a small area of RAM which can be used efficiently for more general purpose data.

Unconditional jumps

One aspect of an assembler which should be examined before committing to it is its handling of jump instructions. Most 4 bit microcontrollers have two types of non-conditional jump - the near jump and the far jump. The near jump is usually a one byte instruction, while the far jump is two bytes. It is therefore wasteful of code space to use far jumps when near jumps will work, as the average program has a lot of jumps in it. The problem with the near jump is its range, usually restricted to the same limited region of code in which the jump instruction resides. When the program is written it is often not known if the target address of a jump is within the range of a near jump. This is particularly true when the near jump is restricted to the same block of addresses as opposed to plus or minus a limited range. With the restriction to block type the target address might only be a few bytes away, but because it is the other side of a block boundary it is inaccessible by the near jump. Which type of jump to put where is then intricately enmeshed with the exact location of each instruction, something which changes every time a modification is made to the code. Having just made a change, the programmer does not necessarily have the information available to know which type of jump to put where, or rather, working out which jumps should change from one type to the other is a tedious and error prone operation, often a waste of time. The assembler, on the other hand, does have this information, so it should be possible to use a single instruction to cover both types of jump, and the assembler should choose a near jump if it can, otherwise put in a far jump. It should be able to do this for forward jumps as well as backward jumps. This may seem obvious but sometimes references to addresses not yet seen are assumed to be in the far range to be on the safe side. A good assembler should look ahead until it has either found the address referenced, or gone to the end of the range of the near jump, thus establishing precisely which kind of jump is required, and so optimising the code. An assembler which does not do this can waste either code space or programmer time trying to optimise the jumps by hand, and having to reoptimise every time any changes are made, and addresses change.

Conditional jumps

As far as conditional jumps are concerned, 4 bit microcontrollers can generally be divided into two categories. One group uses 'branch' instructions, and the other 'skip' instructions. With branch instructions an offset (+/−) from the current position forms part of the instruction, and if the condition is met the branch is followed, otherwise the instruction immediately following the branch instruction is obeyed next. There are usually branch instructions to cover most, if not all, combinations of the various status flags kept by the processor. This is the method that will be most familiar to anyone who has been used to writing assembly language programmes for 8 and 16 bit processors, and it often includes an unconditional branch which is the form the short jump referred to above takes in this type of architecture. There are variations on this theme, for instance in some cases there is one status bit, and one conditional branch, which is obeyed if the status bit is set. There is, accompanying this, a set of 'test' instructions which will set or clear the status bit according to the condition of other flags, such as Carry, Zero, etc. Where this architecture is found, the macro facility comes in useful for making up conditional branches by pairing the 'branch on status set' with whatever condition tests are required. The other common form of architecture found in 4 bit microcontrollers is the 'skip on condition' form. In this, if the condition indicated is true, then the instruction immediately following the 'skip' instruction is ignored, or skipped, and processing continues with the next instruction. Very often the instruction skipped is a jump to be obeyed if the condition is not true. This may seem a clumsy way of implementing conditional branches, but in fact it has various advantages, in that many instructions, not just 'skip on condition' instructions, can be made to skip. Arithmetic operations can be made to skip on carry set – a condition commonly tested in machines with the conditional branch type architecture. Subroutines can be made to skip on return if some condition becomes true during their operation, thus obviating the need for a condition test by the calling section of program. Very often in the 'branch on condition' type, code has to be written to bypass a single instruction. In the 'skip' type this can happen automatically if the condition arising causes a skip in the previous instruction. Clearly there are advantages and disadvantages to both types of architecture and it may only be a case of personal preference as to which to choose. The 'skip' type architecture is generally that where the 'string effect' described below is to be found. In the examples of microcontroller code which follow in the next few chapters, instances of code for both types of architecture will be found.

The string effect

One architectural feature which occurs in machines made by various manufacturers is what NEC call the string effect, and Mitsubishi call 'continuous description'. If an instruction which has this effect is followed immediately by the same instruction, the second one is skipped. For example, LA or LAI (load accumulator) is often endowed with this property, and this allows a subroutine to be written with multiple entry points, each setting a parameter to a different value. Here is an example of how it might be used:

```
ADD4:   LA    4        ;Add two 4-nibble numbers
ADD3:   LA    3        ;Add two 3-nibble numbers
ADD2:   LA    2        ;Add two 2-nibble numbers
        TAB            ;Save count in B register
        .
        .
        .   etc.
```

If calling the routine ADD4, the accumulator takes the value 4, and the two following LA instructions are skipped, the value 4 is transferred to the B register. Calling the routine ADD3 causes the accumulator to take on the value 3, and the LA 2 instruction is skipped, the B register taking the value 3. This saves having to load this parameter prior to calling the routine each time the routine is called, and hence saves space. The same effect could obviously be achieved without the string effect by using jumps, but that is less efficient in its use of code store.

Look up tables

Another technique which is often used in programs to make the best use of the available space in the program store is the use of the look up table. Look up tables can be used in various ways, and some manufacturers have made provision for their use, while others have not. One of the commonest ways of using a table is to provide fixed data for a routine, for instance for supplying the number of days in the month for a calendar routine. The routine which advances the date must check to see if the month has to change, so the month, as a number, is used as an offset in a table, the returned value giving the maximum number of days in the current month. The ease with which this can be done in a given microcontroller depends on the in-built facilities. Several manufacturers include instructions which allow transfer of data from the ROM to internal data registers. Very often the bottom eight bits of the address come from a pair of registers, the remaining bits are specified in the instruction, so are fixed by the assembler. This restricts tables to residing within a single 256 location page of the memory, and care may have to be taken to ensure that this is true, but it is not usually too difficult to arrange. It is often simplest to fix a table with an ORG (origin) instruction, and avoid the need to carry out arithmetic on the table start address in order to find the location of the data given the offset. However, it can be very inconvenient if there are more than one or two tables fixed in this way to fit the operating code round them. A table which is relocatable (usually with the restriction that it does not span a 256 location page boundary), is much easier to work with, but more program steps are required to access the table. Here is an example worked in both ways to highlight the difference:

```
;routine to return the maximum number of days in the month. Month number
;in accumulator on access.

         ORG    800h     ;table starts at page boundary

NDTBL:   DB     31       ;Number of days, 8 bit binary
         DB     28       ;Feb - not leap years
         DB     31       ;Mar
         DB     30       ;Apr
         DB     31       ;May
         DB     30       ;Jun
         DB     31       ;Jul
         DB     31       ;Aug
         DB     30       ;Sep
         DB     31       ;Oct
         DB     30       ;Nov
         DB     31       ;Dec

MAXDAY:  TAY             ;copy month number to Y register
         DCY             ;subtract 1 from Y register
                         ;(As month 1 is at address 0)
         LXI    0        ;Load X register with 0
         LRXA   8        ;Load X and Accumulator from ROM
                         ;address 8XY
         RTS             ;return with data in X and Acc.
```

(Fujitsu MB88540 code, 6 bytes of code, 12 bytes of table, 7 cycles to execute)

A byte of code and a cycle of execution time could be saved by having the table origin at x01h, but that might be even more difficult to organise round.

The same routine with a relocatable table:

```
MAXDAY:  LHLI   TEMP                    ;set mem. pointer to temp. store
         ST                             ;store offset from Acc in TEMP
         LAI    (0Fh AND NDTBL)         -1
                                        ;table start address - 1 (lsn)
         ASC                            ;add offset from TEMP
         JMP    MXD1                    ;jump if no carry
         ST                             ;store least significant nibble
         LAI    ((0F0H AND NDTBL)/16) + 1
                                        ;table start address, bits 4-7,
                                              plus 1 (for carry)
         JMP    MXD2
MXD1:    ST
         LAI    ((0F0H AND NDTBL)/16)
                                        ;table start address, bits 4-7
MXD2:    LAMTL                          ;load Acc and TEMP from table
         RT                             ;return with data in Acc & TEMP

NDTBL:   DB     31                      ;table of values as above
         .
         .
         etc
```

(NEC 75 series code, 14 bytes of code, 11 cycles to execute)

To be fair to NEC the first example could have been written just as easily in NEC code. The extra code and execution time taken up by using a relocatable table might be worthwhile in some applications.

Another common use for a look up table is in connection with the display. Data is often displayed as numbers represented by the 7 segment method, and to translate from the number to be displayed to the pattern of segments to be activated, a look up table is used. In some devices the display control logic automatically performs the translation according to the specified table. All the programmer has to do is indicate which look up table to use by loading its address, or part of its address, into a specific register, and the rest is taken care of in the hardware. There may be restrictions on where in the program memory the look up table can be located, a method which allows for abbreviated addressing. In other cases the display controller associates specific data bits in RAM with each display drive output, and so the programmer has to provide the routine to translate from a number to a segment pattern. This would probably be done with a look up table in a similar way to the example above (see also Chapter 8).

Look up tables are also often used with melody generators. The melodies are coded up in tables specifying the pitch and duration and possibly some other qualities of each note, and when it is required to output a certain melody, the address of its table is loaded into a register and the instruction given to start playing. Again, the location of the tables is usually restricted to allow abbreviated addresses to be used.

Many 4 bit microcontrollers operate slowly compared with larger devices, and have more restricted instruction sets, particularly when it comes to arithmetic functions. Often in a control situation, the control algorithm demands some complicated mathematics to be performed on incoming data in order to generate the required output. In a real time control situation it is often not possible to guarantee that the calculation will be performed in the time available. The solution to this problem which is frequently used is to make use of a look up table. A set of answers is precalculated and the input values used as an index to the table of output values. This is usually several orders of magnitude quicker than performing the calculation, but takes up a lot of space as there are usually a large number of possible answers to store. Sometimes it is possible to use tables to speed up critical parts of a calculation, and this technique is explored further in Chapter 9.

Some manufacturers provide special look up table facilities in order to abbreviate commonly used instructions. For example, loading an 8 bit address into a register pair will normally be a 2 byte instruction, but sometimes it is possible to list a small number of commonly used addresses in a table and then use an abbreviated, single byte, instruction to load the registers with an address from the table. Similarly, commonly used subroutines can be listed in a table so that the call statement can be abbreviated in the case of those subroutines listed, and hence save program space. Such facilities will often place restrictions on the positioning of the tables, and the addressing range of the subroutines called, but they remain useful aids to minimising the space required by the program.

The final aspect of the use of tables to be discussed here is the jump table. Very often there will be a 'JMP @' instruction available, in which the value in the accumulator is added to the program counter to determine which instruction is obeyed next. This is useful for instance when servicing a keypad. The number of the key pressed is loaded into the accumulator and the JMP @ performed. Following the JMP @ instruction is a table of jumps to the various key routines. This is usually much quicker and more efficient than having to perform a series of comparisons and conditional branches. It can also be used as a substitute for a proper look up table if the facilities for such are not available.

It can be seen by the discussion above that the use of look up tables can be a useful tool in writing efficient code, and the look up table facilities provided in a microcontroller need to be examined carefully with a particular application in mind when assessing its suitability for the job.

Subroutines

Most, though not all, 4 bit microcontrollers make provision for the use of subroutines. The way in which this provision is made varies from device to device, some being more restrictive than others. In many machines the method familiar to programmers of larger devices is employed – that of a stack in RAM where the return address is stored on executing the CALL instruction. This has the advantage of allowing subroutine nesting to whatever level may be required, provided there is sufficient room in RAM to accommodate all the stack space needed. The return address, and whatever status bits may also be saved, will probably occupy 4 nibbles of RAM or so, and if RAM space is at a premium this may restrict the way in which subroutines are written and used. Having this type of stack in RAM can be advantageous as there will usually be stack manipulation instructions such as 'POP' and 'PUSH'. These are useful for the temporary storage of other data as well as the return address, and for passing parameters from one part of the program to another.

If there is no general purpose stack like this, there will probably be a special stack for subroutines, and possibly also interrupts, to store return addresses. This will be a fixed length stack, thus limiting the nesting of subroutines, but it will be completely separate from the general purpose RAM, so it has the advantage that no RAM space has to be allowed for the return addresses. The length of this stack will vary from device to device, some allowing up to 8 levels of nesting, others only one, with a few in between. If only one nesting level is allowed it restricts the way in which subroutines are used, as no subroutine can call another. This will probably mean that subroutines have to be split up into shorter sections, and the calling program will have to make more subroutine calls. For example, if the main program were to call a subroutine SUBA, which in turn calls a routine SUBB:

```
        CALL    SUBA
                        SUBA:   .
                                .
                                .
                                CALL    SUBB
                                                SUBB:   .
                                                        .
                                                        .
                                                        RET
                        SUBA1:  .
                                .
                                .
                                RET
                .
                .
                .
                etc.
```

This could obviously be split up into:

```
        CALL    SUBA
                        SUBA:   .
                                .
                                .
                                RET
        CALL    SUBB
                        SUBB:   .
                                .
                                .
                                RET
        CALL    SUBA1
                        SUBA1:  .
                                .
                                .
                                RET
```

The code is a little longer, but not excessively so. The problems arise though when the call in the subroutine is required in a loop. Here it is more difficult to split the code up, and the programmer may have to resort to inserting the code that would have been in the second subroutine several times over in the program, so coding cannot be as compact if the level of subroutine nesting is severely restricted.

Some of the 4 bit microcontrollers available on the market have no provision for calling subroutines at all. They are obviously intended for use in simple applications, but there are likely to be sections of code which are commonly executed, and it may take considerable ingenuity to write compact code for these devices. Some of the techniques applicable here can also be used in devices where subroutines are allowed, but nesting is restricted. One technique is to make use of a table of jumps where the return statement would be. The 'calling' section of code loads a number into a dedicated RAM location prior to jumping to the 'subroutine', and at the end of the routine this number is used as an index to a table of jumps. This means that every use of the routine has to have an entry in the table, and will probably restrict the number of 'calls' to a maximum of 16. Another technique relies on the overall structure of the program being in the form of a loop. As the loop is executed, sections of code in the early part of the loop can set flags, and write parameters to RAM, which sections of code to be obeyed later in the loop can examine, and act on. For example, a flag could be set to indicate that a certain section of code is to be executed, and the data it has to work on can be placed in RAM. This method is only applicable where the results of the 'subroutine' operation are not immediately required. It may be good for such tasks as updating a display, or setting up some external signals, where the response time required is slow compared with the rate at which the software loop is executed. What is possible really depends on what is available in the instruction set in the way of jump instructions, and will be different for every type of device.

Another feature of subroutines worth mentioning, that is available in some 4 bit microcontrollers, is the 'Skip on Return' facility. Normally on return from a subroutine the instruction immediately following the CALL statement is the next one to be obeyed. However, if the instruction set includes a 'Skip on Return' instruction, this can force the instruction immediately after the CALL to be ignored, or skipped, and processing proceeds with the next again instruction. This is particularly useful where a subroutine is examining data for a particular condition, say, for example, the equality of two strings of data. The calling routine sets up the addresses of the strings, and calls the subroutine. The instruction immediately after the CALL is a jump to the section of code to be obeyed if the strings are unequal, the subroutine having a normal return statement in the case of inequality. If the strings are equal, a 'Return and Skip' is obeyed, the JUMP statement is skipped, and processing continues with the code after the JUMP. Often the JUMP is a jump back to the start of a loop, for instance in searching for the occurrence of a matching string. The use of the 'Return and Skip' can often be used to advantage to make code more compact.

7 Software design II – Interrupts, time keeping, time critical routines, power saving techniques and test routines

4 bit microcontrollers are often used in systems where they are required to keep accurate track of the passage of time. This may be in the form of a time of day clock, for equipment that is permanently powered, or timing the duration of some process, the absolute time being irrelevant. There will also be applications where both functions are required, for instance in metering applications where tariffs vary according to the time of day and day of the week – particularly in telephone call cost monitoring. Systems needing full time of day information may not be 'switched on', that is, in use and required to perform their usual control and display functions permanently, but the microcontroller will need power on a permanent basis, probably from a battery. It is with these thoughts in mind that the subjects for this chapter were chosen to go together. One of the major uses of interrupts in 4 bit microcontroller systems is timekeeping of one sort or another, and it is often necessary to perform the interrupt routine in a very short space of time.

Interrupts

As will be seen from the review of devices in chapter 3, most 4 bit microcontrollers support some sort of interrupt system, and those which do not have a full interrupt system have some way of 'waking up' the microcontroller on receipt of some external signal, or perhaps the completion of some internal process which occupied only one of the integrated peripherals. Some form of interrupt handling is usually central to the design of any microcontroller system. The way an interrupt works is like this. Certain events, both external and internal to the microcontroller, are given the capability of setting a flag in the microcontroller internal logic. At the end of each instruction cycle the microcontroller examines the state of this flag, and if it is set it will interrupt the process it is currently performing, save some vital information, such as the address of the next instruction to be performed, and jump to a specially designated routine to take action appropriate to the event causing the interrupt. The events which are allowed to cause interrupts are usually classed into external and internal interrupts. External interrupts are usually one or two in number, and can sometimes be programmed into positive edge, negative edge, positive level or negative level sensitive types. With edge type interrupts the internal flag is set when a voltage transition in the chosen direction occurs, and it is not reset until the interrupt is serviced. With level triggered interrupts, the flag is volatile, and if the interrupt source changes state before the interrupt flag is examined, no interrupt will occur. This may not seem to be an important difference if the flag is examined on every instruction cycle, but in systems where interrupts can be enabled and disabled, and where one interrupt routine can lock out another, it can be some considerable time before the opportunity arises to service an interrupt, and so it is important to make sure that the correct type of interrupt chosen. If the wrong type is chosen, there will be in one case extra code to see if the interrupt is still present and so see if it is necessary to carry out the interrupt service routine, and in the other case interrupts will be lost which should have been acted on if another routine takes longer to process than anticipated, or an unfortunate timing sequence occurs whereby the interrupt is disabled or otherwise locked out just before the critical event occurs.

Internal interrupts are usually triggered by the completion of some task such as a serial I/O line having just completed the transfer of a byte of data, or a timer/counter register having overflowed or counted down to zero. Interrupts are allocated priorities by the microcontroller hardware design. These are usually fixed as far as the user is concerned, and can mean one of two things. In some cases the microcontroller can only service one interrupt at a time, and if this is so, an interrupt service routine cannot be interrupted. The effect of the priority scheme here is to determine which interrupt should be serviced if more than one is active when the interrupt flag is examined. A low priority interrupt can still lock out a high priority interrupt in such a system by virtue of the fact that it arrives first. Note that when an interrupt routine is completed, any other interrupts pending will be serviced in order of priority, so that a low priority interrupt occurring before a high priority interrupt during the execution of another interrupt service routine will not be serviced before the high priority interrupt even though it arrived first. The order of interrupt servicing is not always as simply predicted as it may first appear. In other microcontrollers, multiple levels of interrupt are permitted, so that one interrupt can interrupt another. In this case a lower priority interrupt service routine could be interrupted by a higher priority interrupt, but not vice versa. Such systems usually operate by using the stack in data RAM to store the address to which to return at the end of the interrupt routine, and sufficient space must be allowed in the stack to accommodate all the levels of interrupt.

Another important aspect of interrupt systems in 4 bit microcontrollers is the ability to enable and disable all or selected interrupts as required. The usual arrangement is to have one selective enable/disable command, which selects which interrupts are to be active and another master enable/disable command which allows or disallows all interrupts in the system. Often a design will only require some of the interrupts available in the chosen microcontroller, and so only these will ever be enabled. There will quite likely be times when it is essential that no interrupt can occur, and to cover this section of program, the master disable can be issued. At the end of the critical section, when it is again possible to service any interrupts, the master interrupt enable command is given, but this only enables the interrupts which have been enabled previously using the individual enable command. It is worth considering what happens to any interrupt which occurs while it is disabled. It does not cause an interrupt, but does it set the flag? If it is an edge triggered interrupt the answer is probably 'yes'. Having temporarily disabled the interrupts to allow some time critical section of the program to operate correctly, it is usually necessary at the end of the section to allow any interrupts which may have occurred while it is executing to be dealt with. However, there may be occasions when this is not what is wanted. Take for example an external interrupt line which is connected to a manually operated switch. The switch will probably bounce for some milliseconds when it is operated, and some 'debounce' system is required. This can be handled in software by disabling the interrupts. On the first edge, at the start of the operation of the switch, the interrupt occurs and is serviced. The event which the operator intended to initiate takes place, once. The service routine may be completed before the switch has finished bouncing, and so one part of the interrupt service routine will be to disable any further interrupts from this source. After a predetermined time the software must re-enable the interrupt ready perhaps to detect the switch being released, but if any interrupts occurring while disabled cause the interrupt to be serviced again as soon as it is re-enabled, it could result in the operator initiated event taking place once again, erroneously. The microcontroller should include in its code set some means of clearing the request flag for individual interrupts so that this situation can be catered for. In some processors the interrupt request flags can be polled so that an external interrupt input can be used as a latch to record the occurrence of some external event, but not cause an interrupt. The action of polling the interrupt request flag also clears the flag if it is set, and so offers a mechanism whereby the flag can be cleared if desired, as in the switch bounce situation. In some parts the interrupt enable command takes care of the situation by allowing individual interrupt request flags to be cleared at the same time as the interrupt is enabled.

Interrupt vectors

The usual scheme in 4 bit microcontrollers is a simple vectored interrupt. When an interrupt occurs, the microcontroller, after saving the return address, performs a jump to a location known as the interrupt vector. This is often a fixed address, and a different address is allocated for each type of interrupt. In some microcontrollers however the user can specify the vector addresses. Sometimes two or more interrupts share a vector, and so the interrupt routine, which is the section of code which is executed as a result of the interrupt, will have to begin by establishing which interrupt it is serving. With fixed vectors it is common to place all the vectors for all the different types of interrupt in the same section of code store – if not at adjacent addresses, then not usually more than a few bytes away. The interrupt routine will usually begin with a jump to a more convenient place for the code to be located. The ability to specify the interrupt vectors avoids the need for this, but the facility is usually only available on the more powerful devices.

Interrupt service routines

The first thing to be done in any interrupt service routine, usually, is to save any registers which may be used in the routine, so that they can be restored at the end to allow the interrupted process to be resumed without error. It is important to make sure that all that it is necessary to save is saved. In some microcontrollers the processor status is saved along with the return address. Such things as the carry flag, zero flag and other flags to control the action of the microcontroller are often kept in a 'processor status word', or 'flags register'. If this is saved automatically along with the return address when the interrupt occurs then it will be automatically restored again with the return from interrupt instruction, and the programmer has no need to be concerned about it. However, in some microcontrollers it is up to the programmer to save the processor status along with any other registers which may be changed during execution of the interrupt routine, and to restore them again at the end. Other registers which may have to be saved include memory page or bank registers, if the microcontroller operates in that mode. If the microcontroller has a stack in the data RAM, then that is the usual place to save the registers. It is simply a case of pushing them on to the stack at the start of the routine, and popping them off it again just before returning from interrupt. If there is no stack which can be used in this way, a section of the RAM will have to be allocated to store them in. Provided the architecture is such that only one interrupt can be active at a time there should be no problem in setting aside a section of RAM, and having a standard piece of code which is used at the start of each interrupt routine. In this way the return from interrupt can be one section of code which all the interrupt routines jump to which restores all the saved registers before performing the return from interrupt instruction.

Here is an example from an OKI 6351 microcontroller. In this device the data RAM is addressed by the page and bank mechanism, but the lowest 256 nibbles can also be addressed by a 'Working specification register' technique, whereby one of 16 'pages' is specified, and when memory is accessed, the accessing instruction specifies which of the 16 nibbles in the page is to be addressed. A zero after the nibble number in the instruction indicates the 'working specification' mode rather than the page and bank mode. In this particular software implementation it was decided to use the page and bank method for normal routines, and not to use it for interrupt routines. It was also decided that in normal routines the working specification register would be kept at zero, allowing access in that mode just to the first 16 nibbles of RAM, locations which were used for flags indicating program status, and various temporary stores. This left the interrupt routines free to access the first 256 nibbles of RAM using the working specification register, provided it was set back to zero at the end of the interrupt routines, and avoided the need to save the page and bank registers. The working specification register is set to value 'n' by the instruction 'OUT #n, WORK'.

```
        ORG   400h        ;Timer interrupt vector
        JMP   TIMI        ;jump to timer interrupt routine

        ORG   401h        ;External interrupt vector
        JMP   EXTI        ;jump to external interrupt routine

        .
        .
        .

TIMI:   OUT   #1,WORK     ;Timer Interrupt. Specify page 1 of
                          ;                    data RAM
        MOV   ACC,D0      ;save Accumulator in address D
        INP   FLAG,E0     ;save Flags register in address E
        .
        .
        .

TIMZ:   OUT   #1,WORK     ;end of interrupt routines,
        MOV   D0,ACC      ;restore Accumulator
        OUT   E0,FLAG     ;and Flags register
        OUT   #0,WORK     ;Specify page zero for main program
        RTI               ;Return from interrupt

EXTI:   OUT   #1,IEXM0    ;External Interrupt routine,
                          ;disable further interrupts
        OUT   #1,WORK     ;Save Acc and Flags as in TIMI
        MOV   ACC,D0
        INP   FLAGS,E0
        .
        .
        .

        JMP   TIMZ        ;end of interrupt routine
```

It is usual to keep the amount of processing performed in the interrupt routine to a minimum. There are various reasons for this. Figure 7.1 shows a fairly typical program structure. The main program consists of a loop in which various program status flags are examined, and a set of subroutines which are called when the appropriate flag is set.

The loop is punctuated by a HALT instruction where the program stops until an interrupt occurs. The interrupt service routine takes

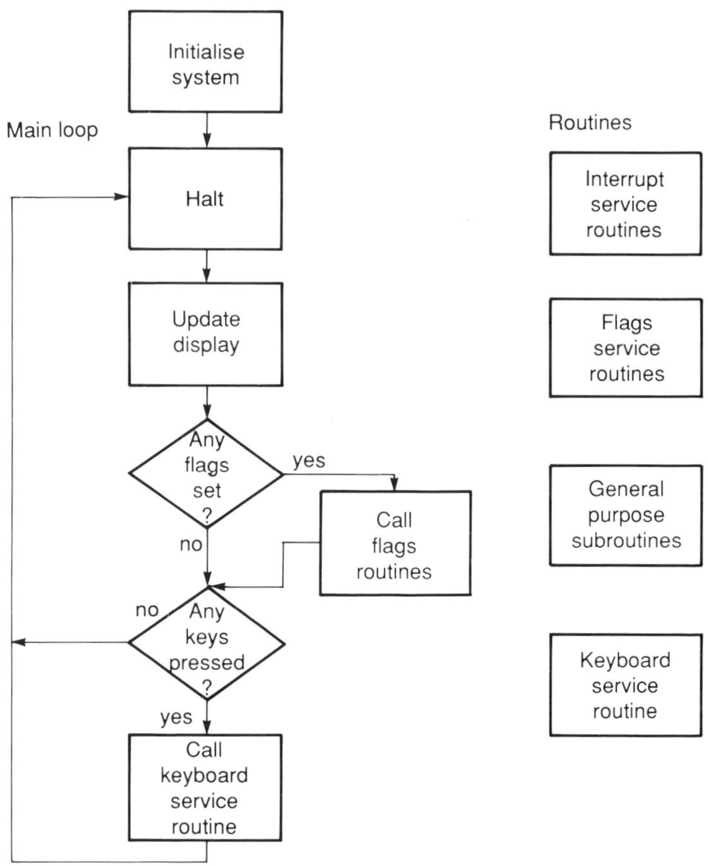

7.1 A typical program structure.

some actions, and sets some flags, and on return from interrupt processing proceeds with the instruction after the HALT. By keeping the amount of work performed in the interrupt service routine to a minimum, it is much easier to keep track of what is happening, and avoid clashes of interest between interrupt service routines and the main program. Suppose for example that there is a subroutine which performs some commonly required operation, such as adding together two 4 digit BCD numbers. This subroutine will include a counter, which may reside in memory. If this routine is to be used from within an interrupt service routine as well as in the main program, it will be necessary for the interrupt routine to save the value of the counter before calling the routine, and restore it again afterwards, so that, should the interrupt occur while the main routine is using that subroutine, the value in the counter is not corrupted as far as the main routine is concerned. The more processing that is performed within the interrupt service routine, the more items like this will have to be saved, using up data memory space, and using more code, possibly unnecessarily. There is also more scope for error, as it is easy to overlook one such item, and it may be a very rare event when the timing is such that the interrupt does cause the corruption. The rarer the event the less likely it is to surface during the development of the product or system, and hence the more likely it is to cause trouble in the field. The alternative is never to allow interrupt service routines to make use of subroutines used by the main program. This is certainly safer, but if it means that subroutines have to be duplicated it is also wasteful of program space.

In some systems it is essential to keep interrupt service routines to a minimum for the system to work at all. If for example, it is required to keep track of time to a fairly fine resolution, it may be necessary to service timer interrupts quite frequently. If the timer interrupt service routine is too long under those conditions the microcontroller will have no time at all to perform its main program. In other systems, it is quite common to have to time the arrival of an external event, often in order to work out a rate of some sort from the period of the incoming pulses. The usual way to do this is for the external event to trigger an interrupt whose service routine stores away the value of some timer/counter register for use by a routine in the main program which carries out the required arithmetic. If the microcontroller can handle only one interrupt at a time it is possible for the external interrupt to be locked out for the duration of an internal interrupt, probably the timer interrupt. The length of the timer interrupt service routine is critical here as it introduces some 'jitter' into the input pulse period measurements, which may be unacceptable if it is too great. It may be possible to minimise the effect by some sort of averaging procedure, but even so the best way to cure the problem is to ensure that the timer interrupt service routine contains an absolute minimum of code, and that any process which can be flagged and then be performed in the main program should be so handled.

Communication between the interrupt service routines and the main program is often by use of flags, single bit indicators, which are used to show that a certain state has been reached. This may be the generation of carry from some counter, or perhaps to indicate some external condition has occurred, on which the main program must now act. There will also be data registers accessed by both main program and interrupt service routines. For example counter registers which are advanced under interrupt and displayed by the main program. When planning a program, and deciding how to make use of data RAM, it is important to make sure that all items common to main program and interrupt service routines are accessible to all. In some microcontrollers this will pose no problem, all of RAM being accessible all the time, but in other devices where this is not so, it may take some careful thought and planning to get it right.

There is another intriguing problem that can arise in the system which has to time the arrival of external events. This comes about when the external interrupt locks out the timer interrupt. The least significant part of the time to be recorded will probably be in a hardware timer/counter register, and when this register overflows it generates a timer interrupt, and under normal conditions the timer interrupt routine is called and the more significant part of the time, held in RAM, is then updated. However, if the external interrupt routine is being processed when the timer interrupt occurs, the time register in RAM will not be updated until the external interrupt service routine has been completed and the timer interrupt service routine can be called. There is a possibility then that the external interrupt service routine will read an incorrect time, reading the hardware register, which has just overflowed, for the least significant part, and the RAM which has not yet been updated for the more significant portion. One possible solution to this is for the external interrupt service routine to examine the value read from the hardware time register, and if it is in the just overflowed state (probably zero), check to see whether the timer interrupt has been serviced. The timer interrupt flag will still be set if the interrupt has not been serviced, and provided this can be examined, appropriate action can be taken. Note though that examining the flag may cancel it, so losing a timer interrupt. This cannot be allowed to happen as it would cause the system to lose time, so it may be necessary for the external interrupt service routine to call the timer interrupt service routine as if it were a subroutine. This may not be possible – it depends on the architecture of the microcontroller in use – and if it is possible it may have implications in stack use or levels of nesting. In some microcontrollers it does work and could be a useful technique.

Non-interrupt systems

In some of the lower end processors there is no interrupt mechanism provided, but it is often still required to respond to asynchronous external events, as well as internally generated conditions, and to have the microcontroller fully active, continuously polling all the conditions of interest so that it can respond with appropriate speed when one arises, would often be far too consumptive of available power. The lower end microcontrollers are often used in battery operated equipment where it is important to minimise power consumption, and the best way to do this is usually to put the microcontroller into a HALT state, where the main processing clock is stopped, and so the CMOS logic stops consuming power. To make provision for this situation the manufacturers allow for certain signals to release the HALT state, and start the processor up at the instruction following the HALT. For example, the inputs from a keypad can sometimes be programmed so that they release the HALT state when a key is pressed, or an internal timer might be programmed to release the HALT state when it times out.

Sometimes, if the processor is at the bottom end of a family of devices, the terminology used for HALT release may be just the same as with interrupts proper, because a common instruction set is used, and the instruction mnemonics are the same. So the instruction that in one processor enables interrupts, may in its little brother enable HALT release. It is important to be aware of this as it can lead to confusion.

Time keeping

Any system which is required to measure time to a reasonable degree of accuracy must be capable of operating with a crystal to set the pace for the oscillator used for timing purposes. As mentioned in Chapter 4, this is usually a 32 kHz crystal, which may or may not be used for the main processing clock as well as the timing clock. A common method of using the 32 kHz crystal is to feed the oscillator output into a programmable divider, and the divided down clock into a counter. The counter generates an interrupt when it overflows and the program takes appropriate action. The counter can often be preloaded, giving the programmer full control over the frequency of the timer interrupts, within the confines of minimum and maximum values imposed by the microcontroller hardware design. In other parts, the 32 kHz feeds into a fixed divider chain, often 15 stages long so that the final output is an accurate 1 second clock. In such a system the programming options will probably allow a selection of taps from this divider chain which will be allowed to generate interrupts. In this way the timer interrupt frequency can be set to any suitable power of 2, from 0 up to 8 probably, and if other frequencies are required to be generated then a counter can be operated by the timer interrupt. It will usually be possible to access the whole divider chain, or at least part of it, as data registers, so external events can be timed with a reasonable degree of precision.

If a simple time of day clock is required, it is just necessary to set aside some spaces in memory for the digits of the time, that is tens and units of seconds, minutes and hours. The numbers can be incremented in the 1 second interrupt routine or perhaps the routine will set a flag to indicate that one second has elapsed, and a subroutine can then be called from the main software loop to perform the time increment function. Chapter 11 gives an example of a system where interrupts are used to keep track of the time of day. If there are any processes in the system which could possibly occupy the microcontroller for more than one second then it will be essential to perform the increment, at least of the seconds, within the interrupt routine. In this case a flag could be set when the minutes were due to change. It may still be necessary to set the 'seconds change' flag as there may be other housekeeping functions which would normally be performed once a second, such as updating the display, or checking the state of some external sensors. Note that the process of incrementing the time requires routines to work to the base ten and also to the base six. Many microcontrollers have instructions specifically geared to base ten, or BCD operations, and working to other bases is usually quite straightforward as will be seen in Chapter 9.

The decision on how frequently the timer is required to interrupt depends on many factors, mostly specific to a given system, but there are some general points to bear in mind. Consideration was given earlier in the chapter to timing external events using interrupts, and here the choice of timer interrupt frequency depends on such factors as the degree of precision required in the timing of events, and the range of counter bits available for reading into the program. Usually in a system there are some inputs to which the system has to respond, but which cannot generate an interrupt. In this case some sort of polling has to be carried out. If the system is operating in a loop which includes a STOP or HALT state, it may be that the STOP or HALT is only released by the interrupt, in which case the timer interrupt frequency also determines the scan frequency for the non-interrupt inputs. As these commonly include a keyboard or other operator interface, it is necessary to ensure that scanning takes place at a suitable rate for the inputs being scanned. In some microcontrollers it is possible to arrange things so that the inputs which do not cause interrupt will release the STOP/HALT condition, and so the scanning frequency is not a consideration when choosing timer interrupt rates. One final point on timer interrupt rates – it may be that a system requires different rates at different times. This is often possible and should be considered, particularly when power consumption is at a premium.

Power saving techniques

In a CMOS system, power is consumed on voltage level transitions. The most generous source of voltage transitions is usually the system clock, so the more parts of the system that are receiving the clock, the more power is consumed. The first place to start looking for power saving therefore is the system clock, where it goes, and what methods are available for limiting it. The slower the clock the fewer transitions will be made, and so the first rule is to use as slow a clock as the system will permit. It will often not be possible to change the clock speed, so the next option is to stop the clock altogether for part of the time. There is often a STOP instruction available to do this, the clock being started again by an interrupt or some other STOP release mechanism. The amount of processing that has to be performed whenever the clock is running also affects the power consumption – the less processing to perform the less power is consumed, so keep the processing to a minimum. This is sometimes in conflict with the requirement to keep the amount of code to a minimum, so a balance may have to be struck between getting all the required code into the available space, and writing the code in such a way as to minimise the number of machine cycles per operation. Other factors are involved here as well, such as time available to complete a given operation. Generally speaking, code which operates in minimum time uses the least power but is not optimal in its use of codespace. In some microcontrollers there are low power modes of operation which shut off the system clock to some parts of the microcontroller, while leaving other sections operating. Make full use of any such operating modes. One point to note when changing between operating modes, especially if an oscillator has been stopped, is that there may be a time penalty. Restarting an oscillator, especially a crystal oscillator, may take a few cycles for oscillation to stabilize, so make sure that due allowance is made for this if necessary.

Apart from looking at the system clock, there are some other ways of reducing power consumption. Often there are external peripheral circuits which are in use only part of the time. If these can be powered down when not in use this may represent a considerable power saving. If there is a display, for instance, it may be possible to blank the display when it is not in use. It is often a simple matter to put a timer into the software which will power down parts of the system which have not been used for a time. With other parts of a system it may be possible to power them down as soon as the function using them is finished. The hardware must of course include the necessary switches to enable the software to operate in this way, all using up valuable I/O. If it is necessary to move to a larger microcontroller to accommodate all the control lines required, the effort may be counter-productive.

Noisy power supplies

Very often a system controlled by a 4 bit microcontroller has to operate in an environment where there is a great deal of disturbance on the incoming mains supply. Much can be done in the power supply design to protect the microcontroller's Vcc or Vdd line, the ultimate being an uninterruptable power supply, with a battery to take over when the mains fails, and sensing and switching circuitry to ensure a smooth switch over from one to the other. This is, however, a costly route, and it may be that the system can tolerate the occasional mains failure, or even short duration dropouts, if the software is designed to take appropriate action.

When power is applied to the microcontroller, there is usually a resistor/capacitor network connected to the reset line which ensures that the controller is held in the reset state until the power has stabilised at the correct operating level. The reset line then goes high, and program operation commences – usually from address zero. In a power failure tolerant system this start up from address zero has to be regarded as a restart after power failure, and any vital system parameters recovered from non-volatile store. The non-volatile store may take the form of a battery backed external RAM, or external EEPROM. There are a number of devices on the market which have interface provision for connecting to an external EEPROM, and there are devices coming on to the market which feature an internal EEPROM, an example of the continuously increasing level of integration that is typical of microcontrollers in general. Provision has to be made for these vital parameters to be stored away just before the power goes off. This can be done using a comparator (many devices have built in comparators) examining a derivative of the raw DC supply prior to the reservoir capacitors, and comparing this with the supply to the microcontroller. The time between the raw DC dipping below a given voltage, and the supply to the microcontroller falling below the specified minimum operating voltage should be reasonably predictable, and it should be possible to ensure sufficient capacity in the reservoir capacitors to allow the software to perform the necessary storage operations. The output from the comparator will need to generate an interrupt, preferably a non-maskable one, so that the 'power fail' routine can be executed as promptly as possible. There is one final consideration – it might be necessary to provide a 'cold start' facility, particularly for starting up the system for the first time, when the vital parameters have not yet been stored, and erroneous operation would result from using the possible random values to be found in the memory. With an external memory it might be possible to preload it with appropriate values, but it may be simpler to have an external signal line which under normal circumstances would be pulled high before the reset pin, but could be held low manually to force a 'cold start'.

There will be systems where power failure to the microcontroller is unacceptable, for instance in a system where the microcontroller is acting as a real time clock. In such systems it will be necessary to provide battery back up provision for the microcontroller, but it may be possible to get away without having to power the whole system from the battery. If a small portion of the system can be identified as essential to retain power, and the rest of the system can be isolated from it in the event of a power failure, it may be possible to make use of a small capacity battery, probably PCB mounted, and save on cost and complexity. Again it will be necessary to provide some detection circuitry so that the microcontroller can be switched into a low power operating mode, and the circuits which isolate the microcontroller from non-powered sections of the system can be operated. In this case there will probably be special routines to be followed when power is restored, and these can be triggered by the power detect mechanism. If the system includes a clock the action taken may be different for different lengths of power failure, and there will be no need to make special provision for the 'cold start' situation.

System tests

A system which includes an intelligent, stored program controller lends itself to being able to perform self-test routines. Most systems specify some degree of self-test, for a variety of reasons. The first is as an aid to manufacture, so that when a system is first put together, correct operation of the system can be quickly verified, and if any faults exist they can be pinpointed easily. The second reason for including self-test is usually to give confidence to the user. If the system performs the self-test routines every time it is powered up, or at regular intervals, the user knows that all sections of the system that can be checked in this way will have been checked regularly, and any faults trapped before they cause possibly expensive damage elsewhere. The third reason for including self-test routines is as an aid to the field service engineer, who has to diagnose problems in a malfunctioning system, and can use all the help available in order to do so as quickly as possible. Some organisations follow the policy that a high quality of self-test routines in their equipment allows them to save on field service engineer training, and also allows them to employ less highly qualified field service engineers.

To build a self-test routine, or set of routines, into a system, to satisfy all these requirements may not be possible, but if the second reason above is to be satisfied some elements of self-test will have to be included as part of the system control software. Giving the user confidence in the equipment, and ensuring that all parts which can be checked in a fully connected system have been checked, are useful functions that can be provided by a subset of the test routines. When the field service engineer has to be called out to a malfunctioning piece of equipment, he or she can switch switches, disconnect motors, plug in special diagnostic connectors, and also replace the microcontroller with a special device containing the extended test routines, provided the microcontroller is fitted in a socket. Such special test devices would be developed in parallel with the principal controlling device, and may be a useful adjunct to the development process, enabling confidence in new hardware designs to be gained.

The nature of any test routines to be built into a system is so varied that it makes little sense to try here to suggest techniques that may be useful. In general, the same techniques that are applicable in driving a system normally are also applicable, with minor modifications, to the test situation. It may be worthwhile though to examine some techniques which can be employed to ensure correct operation of the microcontroller itself. Obviously, in order for the microcontroller to check itself it has to be assumed that it is partially working. In fact it really has to be assumed that the majority of the circuitry is operable for the device to be able to perform the most elementary of checks. Take for example the illumination of an indicator light, driven from a bit on an output port. The light could be put on by a 'stuck at one' fault on the output port, or else it requires that the processor clock is running, the data paths to the program ROM are functioning correctly, the instruction fetch and decode logic is operating, possibly the data will be passed through the arithmetic and logic unit (ALU), almost certainly it will use the accumulator, and it also requires that the data paths to the output port are reliable. To overcome the 'stuck at one' fault problem, some systems use a flashing indicator, which should flash a known number of times whenever the system is powered up. This switching on and off can often be combined with other tests, and hence, if a fault is found, the number of flashes given by the indicator will show how far through the tests it was when the fault was found.

Of the internal sections of the microcontroller the most obvious parts to check are the program ROM and the data RAM. The sumcheck method is the most appropriate to use when checking the ROM. A data word is included in the program whose value is set so that the result of summing the entire contents of the ROM will be zero in the least significant 4 or 8 bits. Doing this is not always possible because of the limited access to ROM data that many microcontrollers give. For instance, in some parts, where there is a 10 bit ROM word, only 8 bits are available to the program, the top two bits determining whether the data is transferred to the internal registers, or to an external port. To try to access program words as if they were data would in this case result in an unpredictable outcome, some of the data being put out on a port which might be performing some control function. Clearly it is not possible to try and test the ROM in this way. One solution is to plant look up tables in separate sections of the ROM, and just check these. In this way it should be possible to ensure that all the ROM address lines are behaving correctly. Note, however, that if the ROM address lines are not correct there is the danger that the data accessed will have that bit set which directs it to an output port. If this method is used, care must be taken to ensure that no damage can be caused by such 'random' data appearing on the output in the event of such a fault.

Checking the RAM should not pose so many problems, as it is designed to be accessed as data. There may be difficulty in accessing the whole of the RAM for testing, particularly if there is a stack in RAM, and subroutines are being used to perform the tests. Test patterns usually applied to RAMs for checking include all 1s, all 0s, alternate 1s and 0s (data equal to both 1010 and 0101), 1s walking in a field of 0s, and 0s walking in a field of 1s. It may not be necessary to perform all these tests, but a high degree of confidence in RAM performance can be gained by performing just one or two of them. Note that these tests also required correct operation of the ALU for comparison of data read from the RAM with what was written.

Sometimes routines written for use in checking the RAM can be useful in other sections of the program as well. For instance, a routine which will write a given data nibble to a block of memory could possibly also be used to clear the display, or put on all display segments. Putting on all the display segments is a common function of power up routines, and allows the user to check that no display segments are missing. Here is an example, in NEC 75xx code, of a routine to clear the memory. This routine starts at the first free location after the stack (which is at the top of RAM) and clears all of memory from there to address 0. The memory pointer is the register pair HL:

```
CLEAR:  LHLI    0     ;Address RAM location 0
        TSPAM         ;Transfer stack pointer (SP) to Acc and
                      ;     memory (at address 0)
        XLDR    0     ;Exchange L and direct memory address 0,
                      ;     puts SP lsn into L
        XAH           ;SP msn to H, 0 to A
CL1:    DLS           ;L = L - 1
        JMP     CL2   ;jump if no borrow
        XAH           ;L = F, so must decrement H
        AISC    15    ;A = A - 1, skip if no borrow
        RT            ;routine complete
        XAH           ;exchange A and H, restores A, puts new
                      ;     value in H
CL2:    ST            ;store 0 at (HL)
        JMP     CL1
```

By presetting A, H and L, and entering the routine at CL1:, other numbers can be stored throughout memory. If a routine is required which stores a given quantity in a limited block of memory it would be possible to include a counter, which could be decremented along with the H register, and so limit the range of the routine.

Mitsubishi, in one of their microcontrollers, have an instruction which transfers data to memory, and then performs an exclusive OR between immediate data and part of the memory pointer. The RAM is 128 nibbles long, requiring a 7 bit address. This is made up from a 4 bit Y register giving the least significant part, a 2 bit X register giving the next part, and a single bit Z register supplying the most significant bit. The instruction 'TAM j' transfers data from the accumulator to the memory at (ZXY), and then performs an exclusive OR with the immediate value j on the X register, putting the result back in the X register. This gives rise to the following routine to fill the memory with a given value. (The stack for the subroutine return addresses is separate from data memory):

```
        SETM0:  LA      0       ;"set memory to zero"
        SETM5   LA      5       ;"set memory to 5"
        SETMA:  LA      A       ;"set memory to A"
        SETMF:  LA      F       ;"set memory to F"
;LA n is a 'continuous description' instruction, see the 'String Effect' above.
                LZ      0       ;address lower half of memory
                BM      STMB    ;call routine STMB
                LZ      1       ;address upper half of memory
        STMB:   LXY     0, 0    ;"Set memory block"
                                ;zero memory pointer
        STMB1:  TAM     1       ;Transfer A to Memory, set X=1
                TAM     3       ;transfer A to Memory, set X=2
                TAM     1       ;A to Memory, X=3
                TAM     3       ;A to Memory, X=0
                INY             ;increment Y, Skip if carry
                B       STMB1   ;branch if no carry
                RT              ;return, all done.
```

The accompanying routine to check the memory contents after filling with a known value can make use of the compare instruction 'SEAM', which compares memory contents with the accumulator, and skips if they are equal. The X register has to be differently handled in this case, so the memory is not read in the same order as it was written.

```
        CKM0:   LA      0       ;"Check memory for 0"
        CKM5:   LA      5       ;"Check memory for 5"
        CKMA:   LA      A       ;"Check memory for A"
        CKMF:   LA      F       ;"Check memory for F"
                LZ      0       ;lower half of memory
                BM      CKMB    ;call CKMB, check block routine
```

```
        LZ      1       ;upper half of memory
CKMB:   LXY     0,0     ;"Check memory block"
        BM      CKMP    ;call CKMP, check page routine
        LXY     1,0     ;address next page
        BM      CKMP
        LXY     2,0
        BM      CKMP
        LXY     3,0
CKMP:   SEAM            ;"Check memory page"
        BL      ERROR   ;branch if A not equal to memory
        INY             ;increment Y
        B       CKMP    ;branch if no carry
        RT              ;return, all correct.
```

There may be a problem here because of the subroutine nesting and the fact that the branch to the error routine comes from within the part of the routine which is sometimes at the third level, sometimes at the second level, and sometimes at the first level of nesting. Probably, if an error is found, very little can be done except light an 'error' indicator, and put the microcontroller into an infinite loop, from which it will only emerge on reset. If this is so, the above code will suffice, but if more complex processing is required the routines above would have to be modified.

8 Software design III – General purpose housekeeping routines

This chapter looks at some of the commonly required functions, to be found in most microcontroller systems, covering such topics as keyboard scanning, display driving, serial I/O, and conversion between binary and binary coded decimal forms. The chapter is concluded with some comments on handling analogue signals.

Flags

Almost all 4 bit microcontrollers are well equipped with bit manipulation instructions. The reason for this is that they are intended for use in situations where there are single bit indicators such as microswitches and other condition sensors to be interrogated, and single bit outputs, such as indicator lights, to be driven. However, this bit manipulation facility is also extremely useful internally, for the manipulation of flags. Flags are single bit indicators used by the software to record the state the system is in. Flags are usually a means of communicating between one part of the program and another, between interrupt service routine and main program, or between one subroutine and another, or even for a subroutine to communicate with itself on subsequent passes. As flags are single bit indicators they can be packed in, four to a nibble, and the bit manipulation facilities of the microcontroller used to examine, set and reset individual flags. Flags are often grouped together in some easily accessible part of the RAM, as they are frequently accessed, and a minimum amount of code should be needed to access them. For example they could be put into bank 0 in a Toshiba system, or be listed in a table for abbreviated reference in an NEC part. Flags will be used extensively in the examples which follow in this chapter, and in Chapters 9 and 11.

Binary to decimal conversion

Very often data is available in binary form, and has to be converted to decimal form for some purpose, particularly when sending it to the display. It may be in binary form for a number of reasons – it may be the result of a calculation which is best performed in binary arithmetic, or it may have been put into binary for storage. A number in the range 0 to 255 takes only 8 bits of storage in binary form, but 12 when coded as decimal. Strictly speaking, only binary numbers can be stored in the types of digital circuit which are covered by this book, so the term 'decimal' should be taken to mean 'binary coded decimal', where four binary digits, one nibble, represent one decimal digit, hence the use of the term 'digit' for nibble by some manufacturers. Here is a method for performing the conversion which can be applied to binary numbers of practically any length. At the heart of the method is the 'decimal adjust' which is available in some microcontrollers. If it is not available it is quite simple to implement in other instructions:- Add 6 to a 4 bit binary number, and if a carry is generated, store the new result. If there is no carry ignore the result of adding 6, and store the original number. After that procedure the decimal equivalent is in the store, with the carry bit representing the next decade. As only one bit, the carry bit, is available here for the next decade, the conversion has to proceed on a bit by bit basis. A flowchart for the procedure is shown in Fig. 8.1, starting with the binary number (B) in the least significant part of a previously cleared section of RAM. For this illustration an 8 bit number is used but the principle applies to numbers of any length. The counters will have to be multi-nibble numbers if there are more than 16 binary bits.

The 20 bits (5 nibbles) of RAM space required by the procedure are mapped out as:

D23 D22 D21 D20 D13 D12 D11 D10 D03 D02 D01 D00 B7 B6 B5 B4 B3 B2 B1 B0

with the most significant bit to the left, and the least significant bit to the right.

Following the flowchart is an example of the routine coded for the Hitachi HD404608 microcontroller. Note that for this device, indirect memory addressing is provided by three registers, W, X, and Y, where W gives the most significant 2 bits of the address, X gives the next 4 bits, and Y the least significant 4 bits. It is assumed in this example that the W and X registers have been set up prior to calling the routine. Another feature of this device is the 'memory register addressing' facility. 16 digits of RAM, from $040 to $04F, can be accessed by special

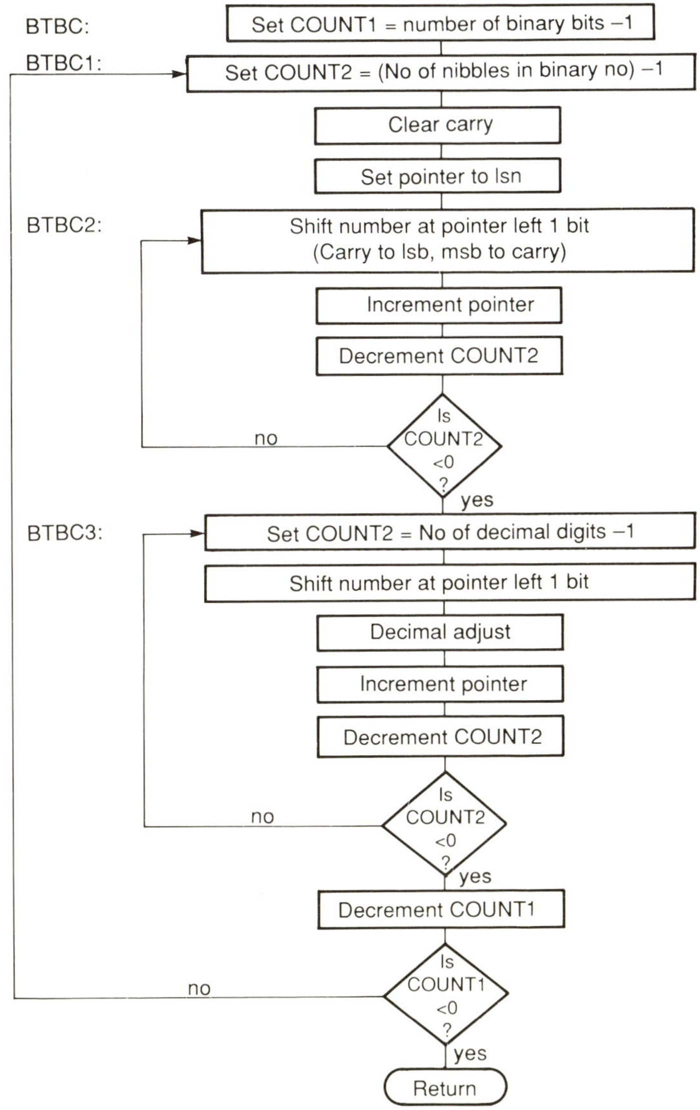

8.1 Binary to decimal conversion.

instructions. They can also, of course, be accessed by normal means, so in the example below, the value of COUNT1 is held in Memory Register 0, which is also addressed as $040 for the add instruction, AMD.

Here is the routine implemented in Hitachi HD404608 code:

```
BTBC:   LAI    7       ;Convert 8 bit binary to 3 digit BCD
BTBC1:  XMRA   0       ;COUNT1 in memory register 0
        LBI    1       ;COUNT2 in register B
        REC            ;reset carry
        LYI    0       ;pointer to lsn of binary
BTBC2:  LAM            ;load binary nibble
        ROTL           ;shift (rotate) left 1 bit
        LMAIY          ;put it back in memory, then increment
                       ;pointer
        DB             ;decrement COUNT2
        PR     BTBC2   ;branch if COUNT2 >=0
        LBI    2       ;set COUNT2 = digit count - 1
BTBC3:  LAM            ;load digit
        AMC            ;add shifts left 1 bit
        DAA            ;decimal adjust
        LMAIY          ;put it back and increment pointer
        DB             ;decrement COUNT2
        BR     BTBC3   ;branch if COUNT2 >=0
        LAI    15      ;15 = -1
        AMD    $040    ;decrement COUNT1
        BR     BTBC1   ;branch if COUNT1 >=0
        RTN            ;Return, conversion complete
```

Without the decimal adjust instruction the routine is slightly longer, as branches have to be put in to take account of carry. For example in NEC 7514 code, the section from BTBC3 would be:

```
BTBC3:  L              ;Load A from Memory
        ACSC           ;Add memory contents plus carry to A
                       ;Skip next instruction if carry results
        JMP    BTBC3A  ;jump if no carry from addition (left shift)
        AISC   6       ;add 6 to A, skip on carry. (There can be no
                                                  carry in this case)
```

```
         JMP    BTBC3B
BTBC3A:  ST             ;store A in memory
         AISC   6       ;add 6 to A, skip on carry
         JMP    BTBC3C  ;jump if no carry
BTBC3B:  ST             ;overwrite memory with new result
         SC             ;set carry
BTBC3C:  ILS            ;increment memory pointer
         etc. . .
```

The counters and loops are used in this routine to minimise the total number of instructions that are required to be stored. To modify the routine for a different binary word length it is necessary to change the loop counters, and if necessary make provision for two-nibble counters. If a faster routine is required, and space is available, the loops can be omitted and the procedure written in straight code.

Decimal (BCD) to binary conversion

A very similar procedure can be used to convert from binary coded decimal to binary form. Again, the operand and the result are stored in adjacent memory, but this time the movement of bits is from left to right, and the decimal adjust is performed by subtracting 3 from any of the decimal digits whose top bit is set after the shift. Figure 8.2 shows a flowchart for this procedure.

Here is the BCD to binary routine in Toshiba TLCS 470 code. The memory pointer is the HL register pair, and H is set before entry to the appropriate value.

```
BCTB:    ST     #7,0    ;COUNT1 at address 0 set to 7
BCTB1:   ST     #2,1    ;COUNT2 at address 1 set to 2
         TEST   CF      ;Clear Carry Flag
         LD     L,#4    ;point to msd
BCTB2:   LD     A,@HL   ;get digit
         RORC           ;Rotate digit 1 bit right
         TEST   A,3     ;Is bit 3 set?
         B      BCTB3   ;Branch if not
         SUBR   A,#3    ;subtract 3 from digit
BCTB3:   ST     A,@HL-  ;put digit back, move pointer
         ADD    1,#15   ;subtract 1 from COUNT2
```

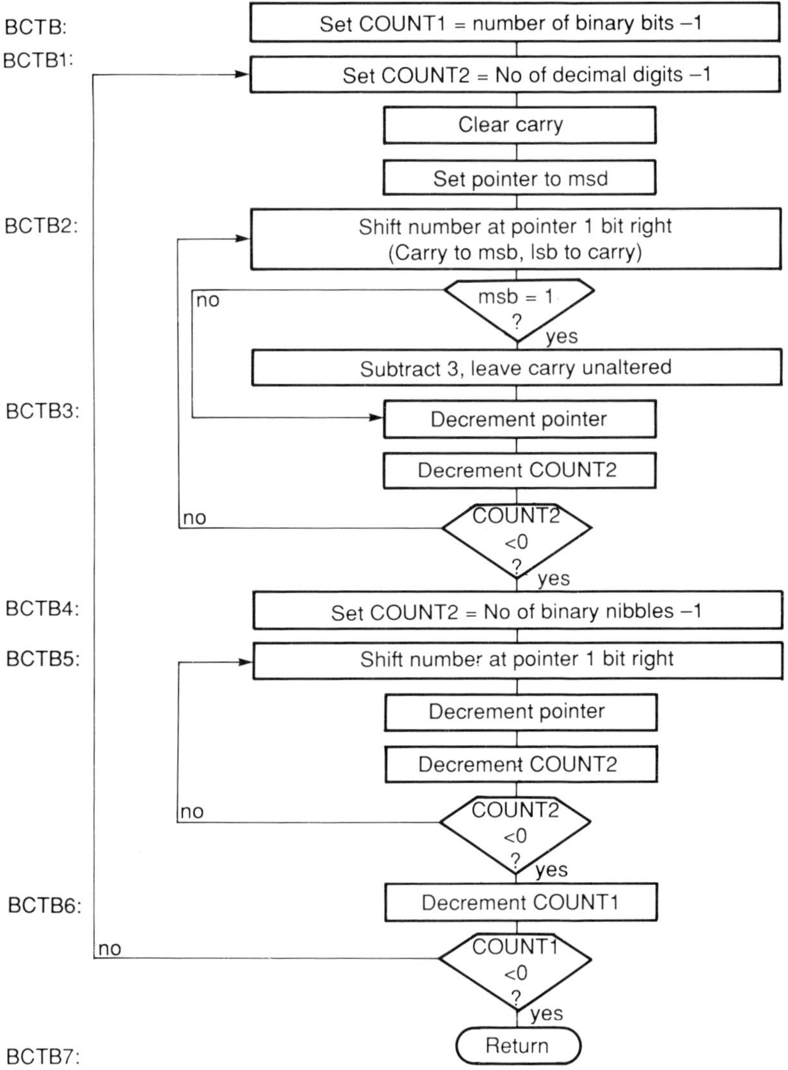

8.2 Decimal to binary conversion.

```
          B      BCTB4    ;branch to BCTB4 if COUNT2 <0
          B      BCTB2    ;otherwise branch to BCTB2
BCTB4:    ST     #1,1     ;COUNT2 set to 1
BCTB5:    LD     A,@HL    ;get nibble
          RORC   A        ;Rotate 1 bit right
          ST     A,@HL-   ;Store shifted nibble, move pointer
          ADD    1,#15    ;subtract 1 from COUNT2
          B      BCTB6    ;Branch to BCTB6 if COUNT2 <0
          B      BCTB5    ;otherwise branch to BCTB5
BCTB6:    ADD    0,#15    ;decrement COUNT1
          B      BCTB7    ;branch to BCTB7 if COUNT1 <0
          B      BCTB1    ;otherwise branch to BCTB1
BCTB7:    RET             ;Return, conversion complete.
```

In this case, because of the nature of the instruction set and the machine architecture, it would be more efficient to set the counters to 16 - bit count, or 16 - digit count. The reason is that the branch instruction is obeyed if the status flat is set, and the ADD y,#k instruction sets the status flag to the inverse of the carry from the add instruction.

Even working from a flow chart these routines may have to be modified to work optimally on the chosen microcontroller.

Keyboard scanning

A keyboard of some sort is frequently used as the primary operator input for many systems. Keys are often dedicated to special functions, and may be arranged in any physical pattern to suit the application, but electrically they can usually be connected into a matrix of some sort. If there are 8 keys or less in the system it is probably simplest to use an input port or two, assigning one input to each key. In systems where more keys are required a matrix arrangement is usually the most efficient in terms of the I/O pins used. For example, 16 keys are simply serviced by a 4 × 4 matrix, using a 4 bit output port and a 4 bit input port, see Fig. 8.3. The input lines are pulled up to the positive supply line, and in order to read the keys each output line is set to 0 in turn, the remaining lines being held at 1. The input lines are then examined, and if any line is 0 it indicates that a key has been pressed.

Figure 8.4 shows a flowchart for a simple routine to scan a 4 × 4 matrix of 16 keys, and return a unique code for each key. There is not room in a single nibble for a code representing 'no key pressed' or

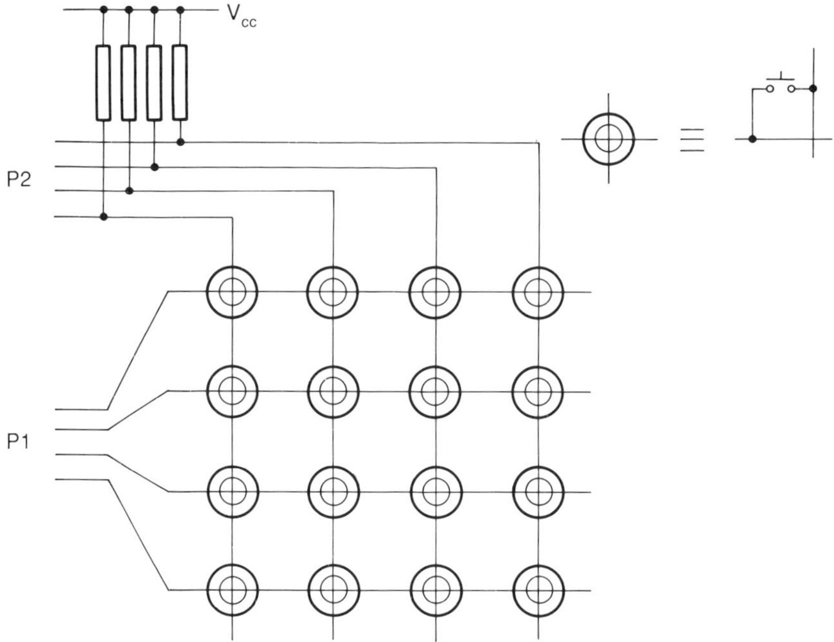

8.3 A simple 16 key matrix.

'invalid keypress', and so a single bit flag, 'Valid Key', is used for this. The 4 bit port driving the row lines is called P1, and the 4 bit port receiving the column lines is called P2. Note that if two or more keys are pressed at once, the 'Valid Key' flag is not set on return from this routine.

Note that it is assumed here that the data output on port P1 is also available to be read in again, using the port as a memory location as well as an output port. If this is not true, as will be the case with some microcontrollers, a separate memory location will have to be used to store the value which has been output to port P1. It is also assumed that the 'rotate right' instruction puts the least significant bit back in at the most significant end, as well as sending it to the carry. Watch out for 'rotate' instructions which put the least significant bit into the carry bit, and put the old value of the carry bit into the most significant bit of the accumulator. If this is the only type of 'rotate' available, carry must be set before the rotate to ensure that only one bit of the port is taken low at a time.

There are various factors which introduce complications to this simple scheme. Pressing two keys at once is not normally permitted,

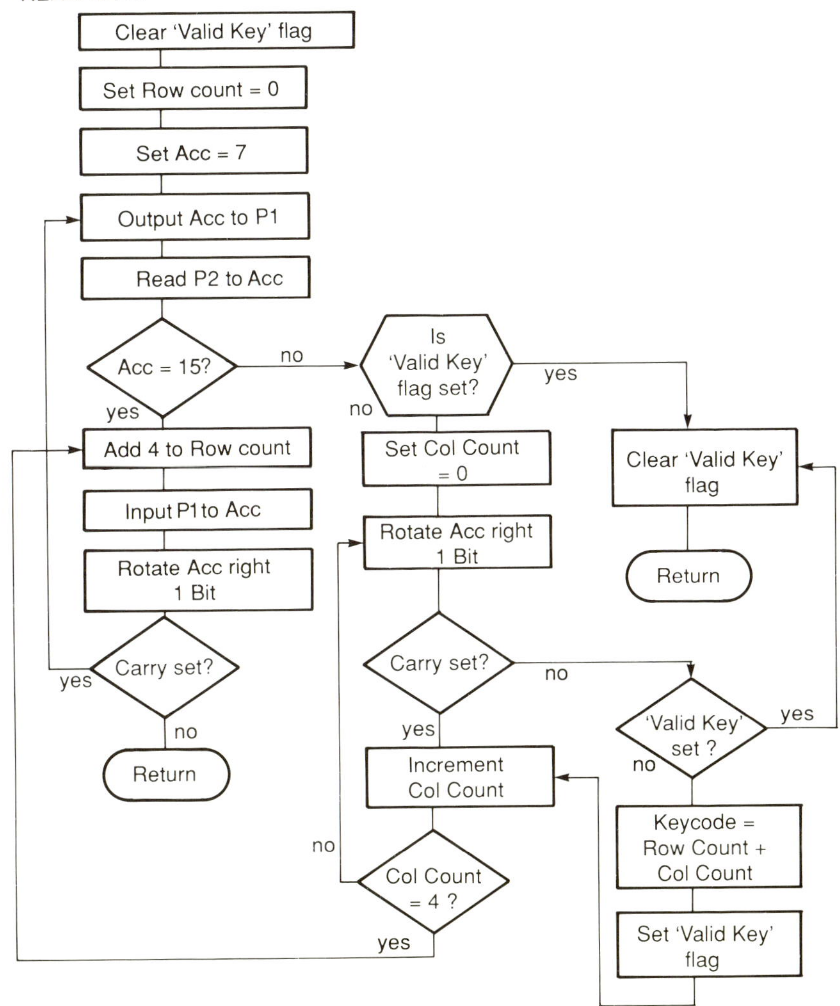

8.4 Scanning a 4 × 4 matrix of 16 keys and returning a unique code for each key.

and the routine above takes care of this, but sometimes it is necessary for two keys to be pressed, for instance if there is a shift key, which may be held down with another key to indicate an alternative function. The second complication is switch bounce. As the operator presses the key and then releases it there may appear to the microcontroller a large number of key presses as the switch contacts vibrate as a result of a single press. Some microcontrollers have special means to handle this but usually it is up to the software to cope with this situation. Other complications can be caused by system specification requirements, for example, is there any special action to be taken if a button is held down for a long time, perhaps an auto repeat function. It may be that it is necessary to take some action when the key is lifted – some item might be displayed as long as the key is held, and erased from the display when it is lifted. The possibilities are endless, and the exact routine required will be different for every system, but here is an example of one key handling routine, see Fig. 8.5. This routine is obeyed once every timer interrupt, at a frequency of 32 Hz. To cope with switch bounce, the routine is short circuited on the pass after a

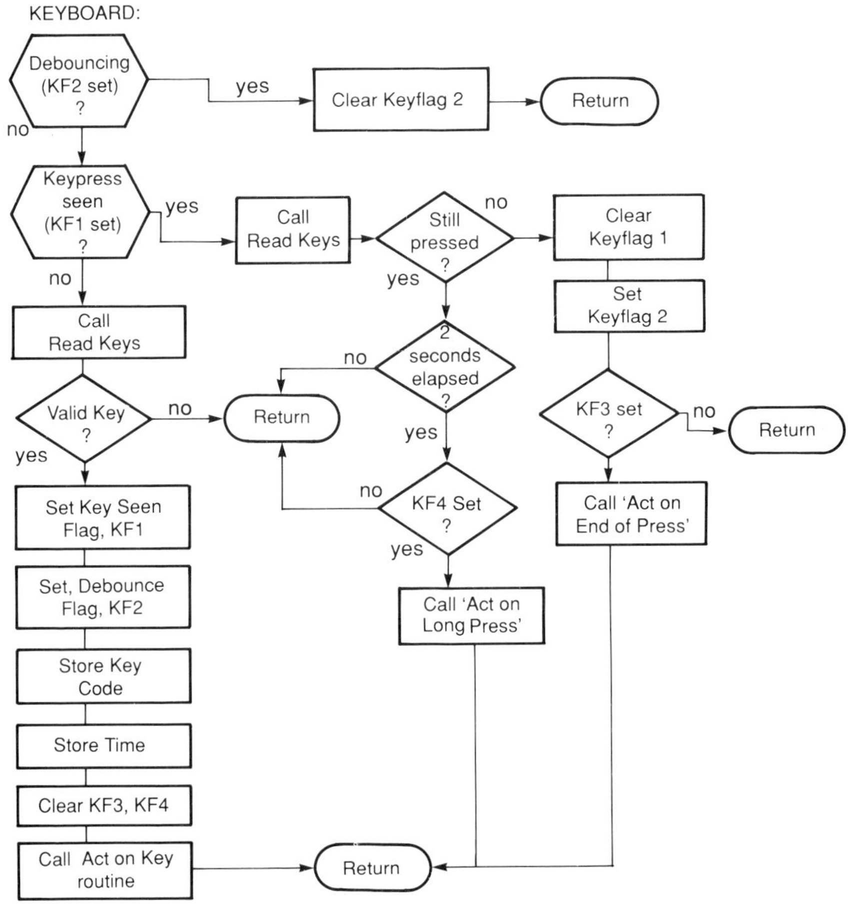

8.5 A key handling routine.

change has been detected. This allows about 30 milliseconds for the keyswitch to settle, which should be adequate for most situations. The routine has four subroutines which it can call, the 'Read Keys' subroutine (shown above) which scans the keyboard, an 'Act on Key' subroutine which is called when a valid keypress is detected, an 'Act On Long Press' subroutine, and an 'Act On End of Press' subroutine. The routine uses a system of flags to keep track of where it is and what is the state of the keys. Keyflag 1 (KF1) is set when a keypress has been detected, and is being acted on. Keyflag 2 (KF2) is set when the keys should be ignored because of possible bouncing. Keyflag 3 (KF3) is set in the 'Act On Key' routine if any action is required at the end of the keypress. Keyflag 4 (KF4) is also set in the 'Act On Key' routine, this time if action is required from an extended keypress.

Driving displays

Chapter 5 introduced the common display types used with 4 bit microcontrollers, and dealt with some of the hardware aspects of their use. Here the topic is picked up again, concentrating on the software aspects. The displays dealt with are the same as in Chapter 5, namely light emitting diodes (LEDs), liquid crystal displays (LCDs) and vacuum fluorescent displays (VFDs).

LED displays

As mentioned in Chapter 5, LED displays are often multiplexed. This makes more efficient use of I/O pins on the microcontroller, and is sometimes essential to drive all the required display elements from the often limited number of high current capacity output lines available. One handy software technique often used to multiplex the display is to have a combined display drive and keyboard scanning routine. The outputs used to pull down each of the keypad row (or column) lines in turn are also used as common lines for each phase of the display. Each line is taken low in turn, and before reading in the keypad data, the display information is output from another port which is connected to the anodes of the LED display elements, see Fig. 8.6.

If the keyboard/display routine is called from a timer interrupt, at say 128 Hz for a 4 phase display, and just one display phase/keypad column is addressed on each pass through the routine, each phase of the display is ensured an equal portion of drive time, and the scanning frequency is high enough to give a flicker free display. The

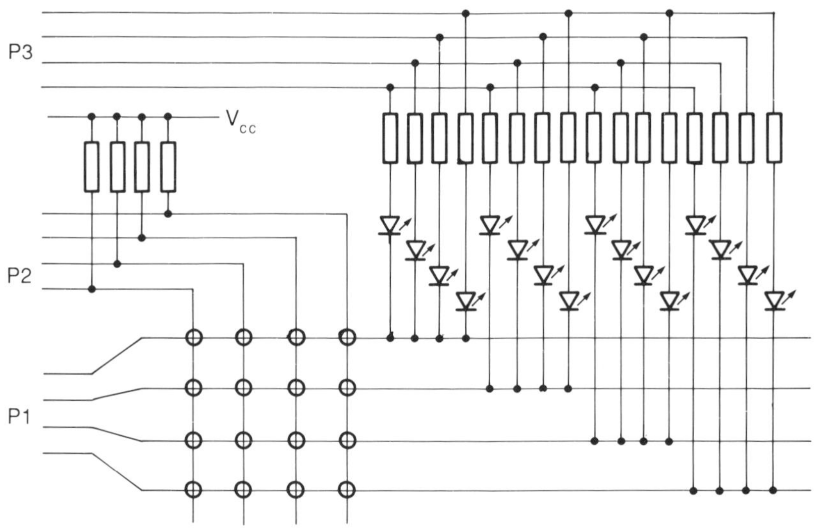

8.6 Combined keypad and multiplexed display drive.

'Readkeys' routine illustrated above clearly needs some modification to cope with this regime. The routine illustrated in Fig. 8.7 is designed to be called with a 128 Hz timer interrupt. Every four times through the routine it sets a flag, 'Keyscan Finished' which is the indication that a routine similar to 'Keyboard' above should be called. This new keyboard routine does not have to call 'Readkeys', as the key code and 'Valid Key' flag have already been set under the interrupt routine. It must, however, clear the 'Keyscan Finished' flag, and the 'Valid Key' flag, set the row count to zero, and also set the row drive nibble (RDN) to 7, ready for the next pass through the interrupt routine. Data for the display is held in RAM, and some indexing method not shown here will have to be employed to access the data and output it to the display drive ports. If there are more than 16 segments of the display this will involve addressing more than one nibble and driving more than one port within this routine.

The routine in Fig. 8.7 organises for the display data to be sent to the multiplexed display, but other parts of the program have to generate the display data and put it into the appropriate area of RAM. Where the display consists of single indicators this is quite straightforward, but often the display will include seven segment characters and the

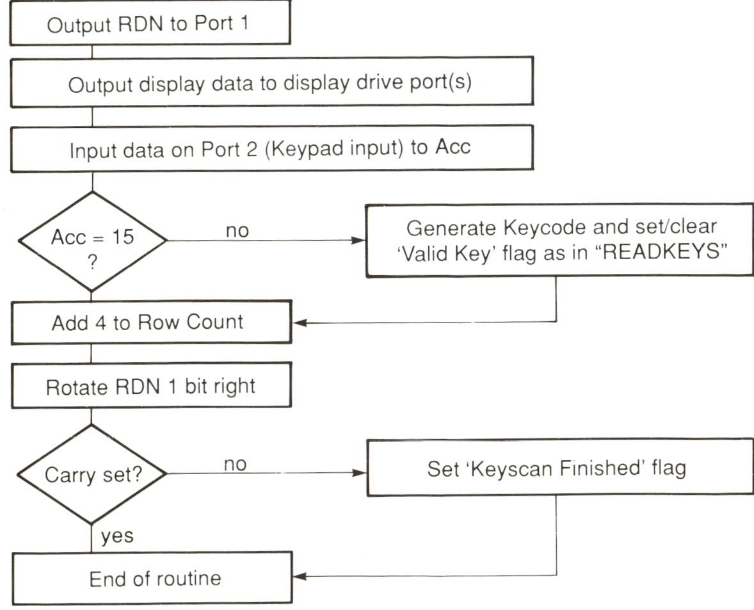

8.7 Flowchart for the routine KEYDISP.

like, and a routine using a look up table will be needed to generate the correct display data. Careful thought may have to be given to the arrangement of the display data in the RAM, and how it corresponds to the segments of the display. It is important, when writing good code, to make sure that only one routine is needed to generate display data from the numbers and other characters which have to be displayed. If care is not taken, it is possible to have several 7–segment numbers on the display, each of which has a different relationship with the display RAM in terms of how each individual segment relates to the common lines, and a different drive routine and look up table is needed for each character on the display. The actual routine for generating the display data could be very similar to the one described below for use with an LCD.

Liquid crystal displays

When using a microcontroller with an integrated LCD driver there are various aspects to consider in the system design. How the display data is to be addressed from within the software routines, and how the display is connected physically to the microcontroller are aspects which can interact, and have to be thought out carefully. The first thing to consider in the programming is how the display is to be initialised. There will usually be several options to choose from, selecting the number of ways the display is multiplexed, the voltage biasing, and the frame rate. There may also be a 'master' on/off bit which has to be set for the display to come on at all. This is useful if the display needs to be switched off without destroying the internally stored display data, or interfering with the display drive software, which can be allowed to continue to update the display data even though nothing is displayed.

An important aspect of the system design is how the display is interconnected with the microcontroller, and this may have an effect on the organisation of the software. In some microcontrollers, the segment drive output pins, and the common backplane drive pins, are dedicated, and setting a particular bit in RAM causes one segment line to be active with one of the common lines. If this is the case it is useful if the system can accommodate any desired arrangement of interconnect between display and microcontroller, for example a printed circuit board which allows tracks to cross over. This then gives greater freedom to arrange the display drive RAM, and how it corresponds with the segments of the display, making it easier to write the display software. If a custom LCD is being used, it also makes life easier for the designer of the custom LCD. In other microcontrollers the relationship between bits in the display RAM and the drive pins is determined by a section of programmed logic on the chip which is defined by the masks used in customising the chip to a particular design. This gives both the microcontroller programmer and the LCD designer complete freedom, and guarantees that it will be possible to interconnect them with a one to one connection pattern. This obviously only applies to mask programmed microcontrollers.

Information on the display usually falls into one of two categories. It is usually either individual indicator segments such as an alarm symbol, a battery low symbol, and such like, or it is alphanumeric data displayed on a seven segment, 14 or 16 segment star, or dot matrix pattern. For the first case, the individual segment indicators, driving is usually easy, just a case of using the bit manipulation instructions to set or clear individual bits in the display RAM. In the second case, where a pattern of segments has to be generated from a numeric code, the task is not so straightforward, and usually requires a look up table. This is where the relationship between the RAM and the display segments has to be carefully arranged, so that each group of segments, for example each pattern of 7 segments, is identically arranged in RAM so that they can be identically driven. Below is an example in NEC code of a seven segment driver. The look up table, which comes first, allows the numbers 0 to 9 to be displayed, together with a blank, and some simple alphabetic characters. The table is located on a page boundary to simplify addressing. The routine itself has two entry points. It is entered at DISNUM when the data to be displayed is pointed to by the register pair DE. Alternatively, if the data is already in the accumulator, A, entry is made at DISNM1. Using the pointer reduces the code required when using this routine several times over to display a multi-digit number. The pointer is unchanged by the routine. On entry to the routine, the register pair HL holds the address of the RAM location where the first nibble of display data is to be written. What the routine does is to take two nibbles of data from the table and write them to the RAM. Using the 'LAMTL' instruction (load A and memory with table data), the more significant nibble of the table data is copied to A, and the less significant nibble to memory, using HL as the memory pointer. The table address used by the LAMTL instruction is placed beforehand into A (msn) and memory (lsn) using the same memory location as will be used for the data, in this case a location called TEMP. The display data represents the seven segments of display, the msn driving segments d,c,b and a (d is the most significant bit), and the lsn driving segments e,g and f, the least significant bit being used elsewhere, see Fig. 8.8. To preserve that least significant bit, the top 3 bits are cleared by the ANL (logical AND) instruction, and the data is written in using the ORL (logical OR) instruction.

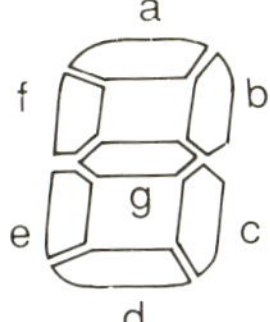

8.8 A seven segment display.

;Display numbers:

```
            ORG     100h
SGTBL:              DB      0FAh    ;0          ;Segment table
                    DB      060h    ;1
                    DB      0BCh    ;2
                    DB      0F4h    ;3
                    DB      066h    ;4
                    DB      0D6h    ;5
                    DB      0DEh    ;6
                    DB      072h    ;7
                    DB      0FEh    ;8
                    DB      0F6h    ;9
                    DB      08Eh    ;A = t
                    DB      03Eh    ;B = P
                    DB      09Ah    ;C = C
                    DB      07Ah    ;D = n
                    DB      01Eh    ;E = F
                    DB      00h     ;F = blank

DISNUM:             LAM     DE      ;DE points to number for display
DISNM1:             PSHHL           ;save segment address
                    LHLI    TEMP    ;address temporary store
                    ST              ;store number to be displayed
                    LAI     0       ;table start address
DISNM2:             LAMTL           ;get pattern from table
                    POPHL           ;restore segment address
                    ST              ;nibble from A to RAM
                    ILS             ;increment pointer for next nibble
                    LAI     1
                    ANL             ;clear all but BIT 0
                    ST              ;put bit 0 back in RAM
                    LADR    TEMP
                    ORL             ;OR in new data from TEMP
                    ST              ;Store it in RAM
                    RT
```

Vacuum fluorescent displays

The facilities available on the various devices capable of driving VFDs vary widely. On the simplest devices there are some output ports capable of withstanding the high voltages necessary in driving VFDs. At the other extreme there are devices which designate an area of memory for the display, and have a hardware display driver which collects data from the memory by DMA transfer, also generating an interrupt to assist with scanning a keyboard, for a combined keyboard/display arrangement similar to the one discussed above for LED displays. Other parts have areas of PLA to provide translation from internal code to segment pattern. The task the software has to perform is therefore highly dependent on the facilities provided in the chip in use. In some parts, where the multiplexing is left to the software, the driving programs will be similar to those described under the section on LED displays above. In other parts the technique will be more akin to the LCD drive routines, where there is hardware in the microcontroller to carry out the multiplexing tasks. One point to watch with VFDs is the problem of ghosting. It is important to put in a sufficient blanking period between driving each section of the display, to ensure that the drive voltage has completely decayed on the undriven lines or else a faint glow of the previous pattern will be left to appear on the next section of the display. There are other causes of ghosting, the commonest being capacitive breakthrough in all off segments, particularly in area displays. This can usually be corrected by adjusting the values of pull down resistors, but this is at the expense of power consumption. VFDs are analogue devices, and consequently the brightness of the display is infinitely variable. Most microcontrollers which provide a control for the brightness will give a number of discrete brightness values to choose from, typically eight.

Software to handle input and output

In the discussion on I/O from the hardware point of view, in Chapter 4, various schemes for passing data between a microcontroller and other circuits were mentioned. The subject divides naturally into serial and parallel communications, serial communications generally being the more taxing on the software. The subject of parallel communications was dealt with in Chapter 4, and as they do not generally require any special treatment in the software they will not be covered further here.

Serial communications

Serial communications using a microcontroller can either make use of the inbuilt hardware facilities for serial data transfer, or it can simply make use of a general purpose I/O port, and build the serial link in software. Whether to do this or not depends on the application, and also on what facilities the microcontroller provides. The usual serial interface provision is a simple eight bit shift register, loaded in parallel from the microcontroller memory, and clocked out by some internally or externally generated clock, for output, or clocked in by a similar clock and read in parallel into memory, for input. If this is adequate for the application, then programming is usually very simple. Often it will use the interrupt generated with completion of an 8 bit transfer either to load another byte for transmission, or read in the received byte. Very often, though, the requirement is for something a little more sophisticated, either using start and stop bits in addition to the data, as in an asynchronous transfer, or perhaps needing more control lines than the simple clock that is available. As an example, here is a routine to store data in an external memory. Data is organised in bytes, but the memory is only one bit wide, as is common in many memory chips. In a large system there would be as many memory chips as bits in the word to be stored, but sometimes the requirement is to minimise chip count, and the one bit wide memory needs fewer I/O pins than the same capacity eight bits wide. In the routine that follows, the least significant four bits of the write address are held in a memory location called WAD0, and has to be output on to Port 3. The data to be written is held in WD0 (lsn) and WD1 (msn). The register pair HL is the memory pointer. The start address for the eight bits to be written is loaded into WAD0 prior to calling this routine, and any other more significant memory address bits can be output on appropriate ports at the start of the routine. Port 6 contains the data on bit 1, the RAM chip enable on bit 2 and the write enable on bit 3:

```
        ;write byte of data to RAM

WRITE0: LEI     7       ;Bit count in register E
WRITE1: LADR    WAD0    ;lsn of address to Acc
        OP      3       ;and then out to Port 3
        LHLI    WD1     ;Address data, msn
        LAM     HL      ;load it into Acc
```

```
            RAR                 ;shift data right, lsb to carry
            ST                  ;and store
            DLS                 ;Address WD0, data lsn
            LAM     HL          ;lsn to Acc
            RAR                 ;lsb of data byte to carry
            ST
            LAI     10          ;activate CE/, data = 1
            SKC                 ;skip if carry set. (Carry is data bit)
            LAI     8           ;data = 0 if no carry
            OP      6
            ANP     6,2         ;Activate Write Enable/
            ORP     6,12        ;CE/ & WE/ inactive
            DES                 ;decrement bit count, skip on borrow
            JMP     WRITE2      ;jump if more to come
            RT                  ;otherwise return.
WRITE2:     IDRS    WAD0        ;Increment address
            JMP     WRITE1      ;write next bit.
```

The routine to read data from the memory is very similar, the data coming in on Port 6, bit 0:

```
        ;read data byte from static RAM

READ:       LEI     7           ;Set bit count in reg E
READ1:      LADR    RAD0        ;get address lsn
            OP      3           ;output it to Port 3
            ANP     6,11        ;activate CS/ (bit 2)
            IP      6           ;collect data (bit 0)
            ORP     6,12        ;CS/ inactive
            RAR                 ;data to carry
            LHLI    RD1         ;address data, msn
            LAM     HL          ;copy to Acc
            RAR                 ;shift data right left and add in new data
                                ;in top bit
            ST                  ;and store it away
            DLS                 ;decrement address pointer to RD0
            LAM     HL
            RAR                 ;shift data right into lsn
            ST
```

```
        DES                   ;decrement bit count, skip if borrow
        JMP     READ2
        RT
READ2:  IDRS    RADO          ;increment address
        JMP     READ1         ;read next bit
```

Asynchronous serial communications

Very often the serial communications are to some standard such as RS232, with a start bit, one or two stop bits, perhaps parity generation, all at some predetermined data rate. Not many microcontrollers have hardware to supply this requirement, and so the work has to be done in software. Usually there is a timer that can be used to time the intervals between individual bits, usually on an interrupt basis, and it may also be possible for the start bit of incoming characters to generate an interrupt. Here is an example of some code designed to send and receive data over an asynchronous serial link at a data rate of 2400 baud.

The received data is input on port 0 bit 0, which is linked to interrupt 4. When a start bit arrives, interrupt 4 is generated. The interrupt routine for interrupt 4 then sets up the timer 1 for receiving the data bits. In this microcontroller there are two timers, and timer 0 is used for data transmission.

```
;Port0.0 input interrupt (int4, vector shared with basic interval timer) occurs
;with start of character rxd from serial link.

p00int:     sktclr    irq4          ;irq4 interrupt?
      ;(skip if true, and clear, interrupt request 4)
            reti                    ;return if not
            skf       flags1.3      ;is this the start bit?
            reti                    ;return if not
            skf       port0.0       ;check for negative edge
            reti
            set1      flags1.3      ;set "character being rxd" flag
            mov       xa,#74h       ;set up timer 1 to time in bits
      ;(16 microsec clock period with 4MHz processor clock)
            mov       tm1,xa        ;timer 1 control register
            mov       xa,#13        ;half bit time 2400 baud
            mov       tmod1,xa      ;timer 1 modulo register
```

```
            set1    tm1.3       ;start timer 1
            ei      iet1        ;enable timer 1 interrupts
            mov     a,#9
            mov     rbcnt,a     ;set up bit count
                                ;(start bit + 8 data bits)
            di      ie4         ;disable Port 0.0
                                ;            interrupts
            reti

;       Timer 1 interrupt routine

;receives a bit on serial i/f and puts it into character buffer when complete
;char rxd, transfers it to message buffer.

t1int:      skt     flags1.3    ;Is Char being rxd?
            reti                ;return if not
            mov     xa,#1Bh     ;full bit time, 2400 baud
            mov     tmod1,xa
            mov1    cy,port0.0  ;read data bit into carry
            clr1    irq4        ;
            mov     a,rchar1
            rorc    a
            mov     rchar1,a    ;roll bit into buffer
            mov     a,rchar0
            rorc    a
            mov     rchar0,a
            mov     a,rbcnt     ;get bit count
            decs    a           ;decrement it
            mov     rbcnt,a
            ske     a,#0        ;char complete?
            reti                ;return if not
            clr1    flags1.3    ;clear "char being rxd" bit
            di      iet1        ;disable timer 1 interrupts
            mov     xa,rxccnt   ;read character count
            adds    xa,#1
            mov     rxccnt,xa   ;put it back incremented
            mov     xa,bufptr   ;message buffer pointer
            mov     hl,xa
            adds    xa,#2       ;add 2 for next time
            mov     bufptr,xa
            mov     xa,rchar0   ;get byte just received
            push    bs          ;save memory bank setting
```

```
            set1    mbe          ;enable address by bank
            sel     mb1          ;select bank 1
            mov     @hl,xa       ;store char in buffer
            pop     bs           ;restore bank setting
            ei      ie4          ;prepare for next start bit
            reti

;Timer 0 interrupt - times bits for transmission via serial link

t0int:      mov     a,(tchar1)   ;get msn of char to tx
            set1    cy           ;for stop bits
            rorc    a            ;rotate carry into msb
            mov     (tchar1),a   ;msn
            mov     a,(tchar0)   ;lsn
            rorc    a            ;lsb of msn to msb of lsn
            mov     (tchar0),a   ;bit to tx now in carry
            mov1    port3.1,cy   ;send bit
            mov     a,(bitcnt)
            decs    a            ;decrement bit count
            mov     (bitcnt),a
            ske     a,#0         ;skip if whole char txd
            reti                 ;otherwise return from interrupt
            mov     xa,tmslng    ;chars left to send
            adds    xa,#0FFh     ;subtract one
            nop                  ;skip this one
            mov     tmslng,xa
            adds    xa,#0FFh     ;test for zero
            br      t0int1       ;branch if buffer empty
            mov     a,#10
            mov     bitcnt,a     ;reset bit count
            skf     flags2.0     ;X-off rxd?
            br      t0int3       ;branch if so
            mov     xa,tbfptr    ;buffer pointer
            mov     hl,xa
            adds    xa,#2        ;add 2 for next time
            mov     tbfptr,xa
            push    bs           ;save memory bank setting
            set1    mbe          ;enable address by bank
            sel     mb1          ;select bank 1 for tx buffer
            mov     xa,@hl       ;get next char
            sel     mb0          ;back to bank 0 for char to tx
```

```
            mov     tchar0,xa       ;
            clr1    port3.1         ;start bit
            set1    tm0.3           ;reset and start timer
            pop     bs              restore bank setting
            reti                    ;return from interrupt

t0int1:     di      iet0            ;stop transmission, buffer empty
            set1    flags2.1        ;more data required
            reti

t0int3:     di      iet0            ;stop transmission if X-OFF rxd
            reti

;*************************************************************************

;           TXCHAR - Start the transmission of chars in buffer
;buffer starts at address 180h (80h in memory bank 1)

txchar      set1    mbe             ;enable address memory by bank
            push    bs              ;save bank selection
            sel     mb0             ;memory bank 0 for pointers etc
            mov     xa,#82h         ;message pointer to 2nd char
            mov     tbfptr,xa
            mov     xa,#08          ;character count
            mov     tmslng,xa
            mov     a,#12           ;init bit count
            mov     (bitcnt),a
            sel     mb1             ;bank 1 for data buffer
            mov     hl,#80h         ;point to first char
            mov     xa,@hl          ;get char
            sel     mb0             ;bank 0 for char being txd
            mov     tchar0,xa       ;store char for interrupt routine
            sel     mb15            ;memory bank 15 for I/O
            mov     xa,#74h         ;select timer mode
            mov     tm0,xa          ;4 microsec clock period
            mov     xa,#104         ;select baud rate (2400)
            mov     tmod0,xa
            clr1    port3.1         ;set start bit
            set1    tm0.3           ;start timer
            ei      iet0            ;enable timer 0 interrupts
            pop     bs              ;restore memory bank selection
            ret
```

This code is for a high end microcontroller, which has some fairly advanced facilities, such as memory mapped I/O, a good selection of memory addressing modes, and easy handling of 8 bit data. In the routines shown above, the receive characters are taken in by the interrupt routines and stored in a buffer. The main program controls where this is by the pointer which the interrupt routine uses, 'bufptr', and is kept informed of how many characters have been received by the character count, 'rxccnt'. Clearly the main routine has to initialise these values before enabling data reception by enabling the port 0.0 interrupts. On the transmission side, the routines shown here take data from a fixed location buffer, and the calling routine has to place data in the buffer and initialise the character count, 'tmslng'. Note that provision is made for X-ON/X-OFF protocol in the transmit routines, in that transmission will be halted at the end of a character if the flag 'flags2.0' is set.

Other routines, not included here, are required to recognise the X-OFF code and set the flag, and also to restart transmission when the X-ON code is received. As there are two separate timers in the chip used for this example, the code has been presented in a way to suggest full duplex operation, that is data transmission and reception simultaneously. Care should be taken though, if implementing such a system – it may not operate correctly. Although the two timers are independent of one another, the reception of a start bit in the middle of the transmission of a data byte could interfere with the transmission timing as one timer interrupt will lock out the other, thus causing it to operate late. How late depends on the speed of the processor, so the interrupt routine execution times must be calculated to see if the delays introduced are acceptable or not. If it is necessary to implement a UART in software it is worthwhile checking to see if the manufacturer of the microcontroller already has the code needed in a library of standard routines.

Analogue signals

A system very often has to deal with analogue signals, either as inputs or outputs, and many modern microcontrollers contain dedicated circuitry to enable these signals to be handled easily.

Analogue to digital conversion

On the input side, when an analogue signal comes into the system it has to be converted to a digital value before it can be used in whatever arithmetic operations might be required by the control algorithm. The commonest architecture found in a 4 bit microcontrollers consists of an analogue multiplexer feeding into a successive approximation converter, where the chosen analogue input is compared with an internally generated analogue voltage which has been derived from a known reference voltage. Several comparisons are carried out, the result of each comparison determining the state of a bit in the output value. Initially the internal voltage is set to half the reference voltage, and the comparison carried out. If the input voltage is above the internal voltage then the most significant bit is set, if not, it is cleared. The internal voltage is then set to a quarter, or three quarters, of the reference voltage (depending on whether the input was above or below the half way point), and the comparison carried out again. This determines the value of the second most significant bit. The process continues in this way until the values of all the bits of the output have been determined. At this point a flag is usually set to indicate that the conversion is complete. In most microcontrollers this process is completely transparent to the user, but it is useful to know what is going on inside the converter in order to make the right decisions when setting up the control parameters which some microcontrollers provide. The usual tasks the program has to perform are selecting which input to use, issuing the instruction to start the conversion, and reading the answer once the conversion has been completed. There may be other tasks to perform, such as selecting the converter speed or accuracy. Most converters operate to 8 bit accuracy, but if a fast conversion is required, and lower accuracy can be tolerated, then it is sometimes possible to specify a 'quick conversion' to say 6 bits. The software may have to determine when the conversion has been completed by polling a status bit, or in some devices an interrupt can be generated on completion of the conversion. If there is no interrupt on completion, but the software requires one, it may be possible to set a timer running when the conversion is started, and allow the timer to generate an interrupt. If the correct time has been set in the timer the conversion will just have been completed when the interrupt occurs.

Sometimes there are restrictions on what can be done, for instance it may be that the same control register is used to select the input and to start the conversion, but it may be necessary to do the two operations separately to allow the comparator input to settle before the first comparison is made. The data sheet for the device in use should indicate if this is necessary, or if input selection and conversion start can be issued in the same instruction. Another restriction sometimes applied is on the use of some, particularly I/O, operations while conversion is taking place. This is to avoid digital noise appearing on the analogue signal being measured, and errors being introduced. This restriction will be a function of the chip layout, and again the manufacturer's data sheet should give the details. It must be read with care. Control and data registers for the analogue to digital converter are usually treated like input or output, often using special port addresses for the various A/D registers.

In some devices, notably the Texas Instruments TSS400 family, the software has to do a bit more than just issue the start command and then wait for the result. In these devices the A/D building blocks are provided, together with special instructions for handling them. The program will have to set values into the conversion output register and read the comparator. This does have the advantage that conversions can be performed more quickly as not all the bits need be tested. It may be known from the characteristics of the input signal that only a limited change in value can take place between conversions, so the most significant bits may not need to be tested. Alternatively, it may only be required to obtain an approximate result, so only as many bits as are required need be tested.

One other point to note about A/D converters is that they are comparatively power hungry bits of circuitry, and may be an embarrassment in a very low power system. If the microcontroller normally operates on a few microamps, but the A/D takes a hundred microamps or so, it may be required by software to switch off the converter when its use is not required. This will be particularly true for equipment operating from miniature batteries. There may be timing restrictions imposed for this - it may be necessary to allow the A/D circuits to settle after being powered up before any accurate conversion can take place. In some parts the powering up and down of the A/D converter will take place automatically, the circuit being switched on with the 'Start Conversion' instruction, and switched off again when conversion is complete. Users of devices which do this should be aware of what is happening as it may cause unexpected variations in supply current demand, and hence unwanted variations to the supply voltage, and fluctuations in display brightness (particularly with LCD displays).

Digital to analogue conversion

The standard technique for digital to analogue conversion available in 4 bit microcontrollers is pulse width modulation (PWM). The reason for this is that it is easy to implement using the types of circuit normally found on such chips, as it requires similar logic to the general purpose timers which most microcontrollers include. The basis of operation of PWM is that a train of pulses can be passed through a low pass filter to give a DC level whose value depends on the mark space ratio of the pulse train. The low pass filter has to be external to the microcontroller, and its characteristics will vary from application to application. The microcontroller will generate the pulse train, and provision is made for the mark space ratio to be varied under program control. The frequency of the pulse train is usually fixed by the CPU clock, though there may be a choice of division ratios. The program has to load a register with a value representing the length of the 'mark' portion of the output waveform, the length of the 'space' portion then being determined by the difference between the 'mark' and the basic period of the output. It is common to use a 14 bit counter in the timer, in which case the basic period is $f_c \times 2^{14}$, and if a value m is specified for the 'mark' time, the output voltage will be given by:

$$V_{out} = V_{ref} \times m/2^{14}$$

This assumes that the rectangular waveform at the input to the low pass filter is switching between V_{ref} and 0 volts, and that the low pass filter has a suitable frequency characteristic compared with the frequency of the waveform generated by the microcontroller. The program has to calculate or look up the value for m, and then place it in the appropriate register, a multi-nibble operation. Arriving at the required value for m depends entirely on the application. Driving a motor will probably involve some sort of feedback loop from a speed sensor, and some suitable control algorithm to ensure the desired speed is maintained under all permissible load conditions. Controlling a tuner would probably usually involve extracting a value from a look up table in order to select a channel chosen by the operator, but it may also involve a step by step change in some sort of 'search' operation. Even in the case of the look up table there may be a measure of feedback, and consequential voltage adjustment in order to provide an automatic fine tune facility.

There may be restrictions on the range of values permissible for m, particularly the minimum value, preventing the output from reaching a true zero. It will probably be possible to get to within 1 or 2% of V_{ref} and of zero volts.

It is usually possible to set up the timer circuit to provide an interrupt at the end of each cycle so that any revision of the value of m can be made at the start of the cycle, during the minimum marking time of the output, so that any alterations in the output can be made in synchronism with the waveform, and not introduce any peculiar irregularities in the waveform.

9 Software design IV – Maths routines

Most systems using microcontrollers require a certain amount of maths, even if it is just addition and subtraction. So when limited to operating with 4 bit words, and using a limited instruction set, the simple formula or algorithm that has to be coded suddenly looks much more complex, and difficult to perform in the required time. This chapter examines some possible ways of handling multi-nibble operations, including add, subtract, compare, multiply and divide, mainly in binary, but addition and subtraction will also be treated in other bases. The chapter also includes a section on using logarithms to get quicker results where there is room for a small percentage of error. It is not the aim here to examine algorithms for performing the more advanced functions that are required, say, in a scientific calculator. The values of such functions can usually be obtained by summing a series whose individual members are evaluated by the simple add, subtract, multiply and divide rules. The routines described here deal only with unsigned integers, but the principles can be extended to deal with other cases.

Multi-nibble addition

The basis of multi-nibble addition is quite straightforward. The corresponding nibbles of each of two operands are added, together with the carry bit from the addition of the previous pair, and the carry is generated which contributes to the addition of the next pair.

The implementation of such a scheme depends on the instructions available in the chosen microcontroller. If there are two address registers or data pointers it is quite easy. The pointers are set by the calling routine to the least significant nibbles of the two operands, and the addition routine has to perform the addition, increment the counters, and count the nibbles, as shown in the flowchart, Fig. 9.1.

On return from this routine the sum of the two operands has

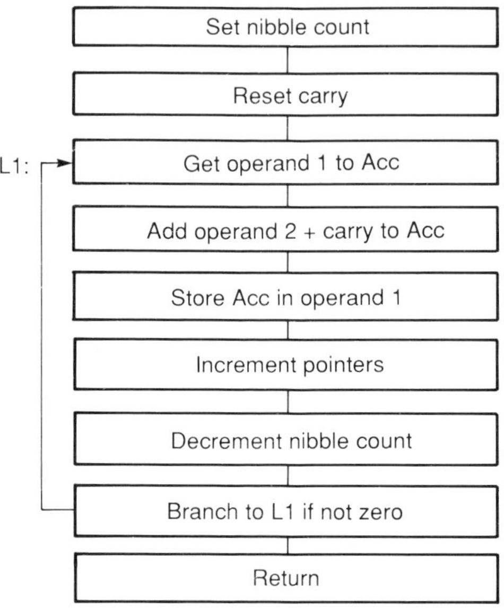

9.1 Flowchart for a multi-nibble addition routine.

replaced operand 1, and carry is valid. This requires that it is possible to decrement the nibble count without disturbing the carry bit – this is usually possible, but it is best to look carefully at the operation of the microcontroller in use. Here is the routine coded up for an NEC chip (the data pointers are DE and HL):

```
ADD4:  LAI    3        ;nibble count - 1
       XADR   COUNT    ;store nibble count
       RC              ;clear carry
ADD0:  LAM    DE       ;get first operand
       ACSC            ;Acc = Acc + (HL) + carry
       NOP             ;skipped if carry
       XAM    HL+      ;Store result, increment pointer
       IES             ;increment other pointer
       DDRS   COUNT    ;decrement nibble count
       JCP    ADD0     ;jump if no borrow
       RT              ;otherwise return
```

Note that the data of each operand is stored in consecutive addresses, and that only the lower nibble of each pointer is incremented, so the data must not span a 16-nibble boundary.

If the microcontroller in use is one of those (and there are several of them) where true indirect addressing is not possible, the approach will have to be slightly different. This is also the case if there is only one pointer available, as for example in some of the Sanyo parts. Here a possible approach is to copy one operand into a common working area (using a subroutine for the purpose), and then set the data pointer to the other operand and call the adding routine, which then adds the data in the common working area to the data indicated by the data pointer. The code for such a technique is shown below. In this example the data is stored in a slightly unconventional way, in that the data nibbles are not stored in consecutive addresses, but at intervals of 16. This allows the routine which increments the data pointer to operate only on the upper nibble of the two nibble pointer. It also makes it easier to use the common working area, as the addresses here consist of a nibble of immediate data which specifies the lsn of the address, and a nibble in a register called the bank register which specifies the msn of the address. By using a constant number for the lsn, and incrementing the bank register, the routine can be written in a loop, and the same routine used for different length numbers by changing the nibble count. One further point to note about this particular group of Sanyo parts is that the indirect memory addressing, making use of the data pointer, only takes place when two conditions are met at the same time. These are that the appropriate enabling flag is set, and also that the immediate address specified in the instruction does not match, in its msn, the value currently held in the bank register.

First the copy routine. The calling program sets the data pointer to indicate the data to be copied into the common working area:

```
COPY 4: LDS     COUNT, 4      ;Set nibble count to 4
        SBNK    1             ;Set Bank Register to 1
        SF1     2             ;Enable the use of the Data Pointer
CP41:   MRW     0,0           ;Using address 0 here
                              ;forces indirect address.
;MRW x, w instruction copies data at chosen address to a nibble
;addressed by the Bank Register (msn) and 8 + w.
        CALL    INCDB         ;Increment pointers and decrement COUNT.
        BANZ    CP41          ;branch if Acc not zero
;          (routine INCDB sets Acc to zero when COUNT = 0.)
```

```
        RTS                   ;return to calling program
```

Now the routine which performs the addition. Again the Data Pointer is set by the calling program. The result overwrites the data specified here.

```
ADD4:   LDS     COUNT,4       ;set nibble count
        SBNK    1             ;Bank Register addresses common
                              ;            working area
        SF1     2             ;enable Data Pointer
        RCF                   ;reset carry flag
AD41:   ADCI    8,0           ;Add carry + 0 to data in common
                              ;            working area
        ADC*    0             ;Add Acc + carry + data at pointer,
                              ;          store at pointer
        CALL    INCDB         ;Increment pointers and
                              ;decrement COUNT
        BANZ    AD41          ;INDCB sets Acc to 0 if
                              ;COUNT = 0
        RTS                   ;return.
```

Finally, here is the routine INCDB, which increments the pointers and decrements the counter:

```
INCDB:  MBNK    TEMP          ;store value of Bank Reg.
        MDPR    TEMP1
        MBR     TEMP1         ;store value of Data Pointer (msn)
        SBNK    0             ;Bank Register set to 0
        ADNI*   TEMP,1        ;Add 1 to TEMP, carry flag not changed
        ADNI*   TEMP1,1       ;add 1 to TEMP1
        ADNI*   COUNT,0Fh     ;Subtract 1 from COUNT
        BAZ     IDB1          ;Branch if Zero
        MRDH    TEMP1         ;New value to Data Pointer
        MRBK    TEMP          ;New value to Bank Register
        ORI     TEMP,0Fh      ;Set Acc to 0Fh
        RTS
IDB1:   ANDI    TEMP,0        ;Set Acc to zero
        RTS
```

These routines are more easily understood if read in conjunction with the memory map:

lsn msn	0	1	2	3	4	5	6	7	8	9	A...
0	TEMP	TEMP1	COUNT								
1	OP10	OP20							WK0		
2	OP11	OP21							WK1		
3	OP12	OP22							WK2		
4	OP13	OP23							WK3		
.											
.											
.											

The locations WK0–3 are the four nibbles of the common working area, and OP10-3, OP20-3, are the two operands being used. The operands do not, of course, have to be located where they are shown above, but each operand does have to be arranged with its nibbles in the pattern shown.

Subtraction

If the microcontroller includes subtract instructions, the subtraction routine can be almost identical to the addition routine, substituting the subtraction instruction for the add instruction. The carry, of course, now represents 'borrow', but retains its validity. If there is no subtract instruction, it will be necessary to perform the subtraction by an invert and add process. To subtract one number, A, from another number, B, in binary is the same as to add the 2's complement of A to the number B. The 2's complement of A is 1 + (1's complement of A), or 1 + (inverse of A). Most microcontrollers have some means of inverting the contents of the accumulator, whether directly by a 'Complement A' instruction, or less directly by an exclusive OR with 0Fh. The 1 that has to be added to make the 2's complement can come from the carry, which becomes an inverse borrow. For an example, take the Sharp SM590, a low end part with only 41 instructions in its instruction set. The RAM, only 32 nibbles in all, is addressed by the B register, consisting of a four bit BL register containing the lower 4 bits of the address, and a one bit BM register containing the 5th bit of the address. For this example it is assumed that the two operands lie in the same half of the RAM, so that BM can be left constant. As there is only one memory pointer, use is made of the X register, a general purpose 4 bit register, for one of the addresses. On entry to the routine the lower part of the address of the least significant nibble of the subtrahend (the number to be subtracted) is in the BL register, and the lower part of the least significant nibble of the address of the minuend (the number from which the subtrahend is to be subtracted) is in the X register. The difference overwrites the subtrahend. Note that in this routine the count has been dispensed with by placing one of the operands, in this case the subtrahend, at the top of the half of memory, so that incrementing the address beyond the end of the operand generates a carry and causes the jump instruction to be skipped. (The carry flag is not affected by this.) The routine is easier to follow if the movements of the data and addresses between the BL, X, and Acc registers are understood. The table below shows these movements, using D for data, P1 for subtrahend address, and P2 for minuend address.

	X	A	BL	
after 1st LDA	P2	D	P1	
EXAX	D	P2	P1	Exchange A & X
XBLA	D	P1	P2	Exchange BL & A
EXAX	P1	D	P2	
EXAX	D	P1	P2	
XBLA	D	P2	P1	
EXAX	P2	D	P1	

So here is the routine:

```
SUB:   SC          ;Set carry (not borrow)
SB1:   LDA         ;Load subtrahend nibble into Acc
       EXAX        ;Swap D & P2
       XBLA        ;Swap P2 & P1
       EXAX        ;P1 to X, Data to Acc
       COMA        ;Complement Acc
       ADC         ;Acc & minuend nibble + carry
```

```
        INBL                ;Increment minuend address
        EXAX
        XBLA
        EXAX                ;Swap addresses
        STR                 ;store difference nibble
        INBL                ;Increment subtrahend address
                            ;skip next instruction if carry
        TR        SB1       ;Jump to SB1
        RTN                 ;Return.
```

Comparisons

Very often it is required to know which is the greater of two numbers, without actually calculating the difference between them. A subroutine to compare two 4 nibble numbers is used by the division routine shown below. The routine to compare two numbers starts at the most significant end, comparing nibble by nibble. As soon as two non-equal nibbles are found the routine can stop. Some microcontrollers have a 'Return and Skip' instruction, which allows a subroutine to return to the instruction but one after the CALL, thus avoiding the need to test the condition after calling the compare routine. If this is not available the carry flag will usually serve to indicate the result of the compare. Strictly speaking there can be three results from the comparison: $A < B$, $A > B$, and $A = B$. It is usually enough to allow the equality condition to be associated with either of the two inequalities, but care has to be taken when calling the routine to ensure that the operators are chosen the correct way, bearing in mind which condition includes the equality case. The code example below is written for a Hitachi microcontroller, which has a compare in its instruction set. The architecture also features a double set of memory pointers - the X and Y registers, both 4 bits wide, are backed up by the SPX and SPY registers. The instruction LAMXY loads A from the memory and swaps X with SPX and Y with SPY, thus changing the memory pointer. Before this routine is called, X,Y must be loaded with the address of the most significant nibble of the 1st operand, and SPX,SPY loaded with the address of the most significant nibble of the 2nd operand. There is also a B register in this microcontroller which is used here as a counter.

```
CMP4:   LBI     3       ;load counter
CP41:   LAMXY           ;get 1st operand, address 2nd
        ALEM            ;compare A with Memory
                        ;set status bit if A <= Mem
        BR      CP42    ;Branch if status bit set
        REC             ;Reset Carry
        RTN             ;return with carry not set,
                        ;1st operand > 2nd operand
CP42:   DY              ;decrement 2nd operand address
        XSPXY           ;swap addresses
        DY              ;decrement 1st operand address
        DB              ;decrement count, set status if no borrow
        BR      CP41    ;branch if status set
        SEC             ;set carry
        RTN             ;return with carry set,
                        ;1st operand <= 2nd operand
```

Note that, since only the Y register of the addresses is decremented, the operands must not cross a 16 nibble boundary. Also, since only the lower 8 bits of address (there can be 10, the top two being held in the M register) are swapped, the two operands must both reside within the same 256 nibble page of memory. The calling program must perform a branch on the state of the carry bit after calling this routine.

Multiplication

The functions so far considered in this chapter have assumed that the operands, and the result of the calculation, will all occupy the same wordlength. Obviously, with addition, the result can be longer than the two original operands, but this is taken care of by the carry bit. However, with multiplication, the product of two numbers can have a total bit length equal to the sum of the bit lengths of the input operands, thus the product of two 16 bit numbers, 4 nibbles, can occupy 32 bits, 8 nibbles. In view of this it is often required to multiply together two numbers of unequal bit length, for example a 4 bit number by a 16 bit number. In such a case the result is frequently required to fit into 16 bits, and so the range of numbers used as operands will be restricted. These are often the result of measurements or entered data and consequently should be subjected to validity checking to ensure that they lie in a valid range. Often the multiplication is used to prescale such data prior to further calculation. The example given here, of a 4 bit by 16 bit multiplication to give a 16 bit result, shows the general principle which can be extended to provide any required combination of bit lengths.

The basic method employed here is just the same as performing decimal long multiplication by hand, except that, since the numbers are in binary, it is just a case of adding in one times the multiplicand at each stage if a 1 is present in the multiplier, or adding nothing if a 0 is present. In between each stage of addition the multiplicand is shifted up one place (multiplied by 2), equivalent to keeping the numbers in the right columns in long hand multiplication:

```
MUL:    XADR    MULT        ;save multiplier
        LAI     3           ;multiplier bit count - 1
        XADR    COUNT
ML1:    LDEI    MLPC        ;point to multiplicand
        LHLI    ANSW        ;point to answer
        LADR    MULT        ;get multiplier
        RAR                 ;lsb to carry
        XADR    MULT        ;store shifted version
        SKC                 ;skip if addition required
        JMP     ML2         ;jump if no addition
        CALL    ADD4        ;add multiplicand to answer
ML2:    DDRS    COUNT       ;decrement count, skip on borrow
        JMP     ML3         ;no borrow, not done yet.
        RET                 ;return, all bits done
ML3:    LHLI    MLPC
        LDEI    MLPC
        CALL    ADD4        ;double multiplicand
        JMP     ML1
```

Where ADD4 is the first of the addition routines given above. No check is made here for carry after the ADD routine. This would represent an error, assumed to be guarded against by range checking the numbers before entry to the routine. It is also assumed that the answer space is cleared beforehand. This routine starts with the least significant bit and works up, the way sums are usually done. That suits this particular instruction set which has no ROL (rotate left) instruction. In some cases it might be more efficient to work the other way, particularly if the multiplier is a multi-nibble number which is much smaller than its field allows. The routine can start with an abbreviated loop, shifting the multiplier left until the most significant 1 is found, which saves a lot of unnecessary doubling of the multiplicand – though if starting at the most significant end, the result, rather than the multiplicand, has to be doubled at each stage. Both multiplication and division can be time consuming operations, especially in the lower end microcontrollers where the cycle time is relatively long. It is possible to speed things up a bit by doing away with the loops and writing the process out 'long hand'. This is repetitive, and wasteful of code space, but if the space is otherwise going to be unused it may be worth doing. It will save on counters and jumps, but it will not make a dramatic difference to the time spent on the operation. If time really is at a premium, and an exact answer is not required, it may be possible to use a logarithmic method described below.

Division

Division, like multiplication, can be performed in a way that is similar to long decimal division, only it is simpler. With decimal division there is the problem of deciding at each stage how many times the divisor will go into the portion of dividend under consideration. With binary arithmetic there is no such problem. The divisor is first shifted up so that its most significant 1 is in the most significant position of the field. It is then compared with the dividend. If the dividend is greater than or equal to the divisor, a 1 is put into the quotient, and the divisor subtracted from the dividend. If the dividend is less than the divisor a 0 is put into the quotient. The 1 or 0 is put into the least significant bit of the quotient, and then the quotient is doubled and the divisor halved until the divisor is back to where it started. This can be coded as follows:

```
DIV:    LAI     0
        XADR    COUNT       ;clear count
DV1:    LHLI    DIVIS       ;address msn of divisor
        SKMBF   3           ;examine msb, skip if 0
        JMP     DV2         ;jump if msb = 1
        LDEI    DIVIS
```

```
        CALL   ADD4     ;double divisor
        IDRS   COUNT    ;increment count, skip if carry
        JMP    DV1      ;
        RTS             ;return and skip if dividing by 0
                        ;(error return)
DV2:    LDEI   DIVIS    ;address divisor
        LHLI   DIVID    ;address dividend
        CALL   CMP4     ;compare, skip on return
                        ;if dividend >= divisor
        JMP    DV3      ;jump if divisor > dividend
        LHLI   DIVID    ;address dividend
        LDEI   DIVIS    ;address divisor
        CALL   SUB4     ;subtract divisor from dividend
        LHLI   QUOT     ;address quotient
        SMB    0        ;set lsb = 1
DV3     DDRS   COUNT    ;decrement count
        JMP    DV4      ;jump if no borrow
        RET             ;routine finished
DV4     LDEI   QUOT
        LHLI   QUOT
        CALL   ADD4     ;double quotient
        LHLI   DIVIS
        CALL   SHR4     ;halve divisor
        JMP    DV2
```

This routine has been written for 16 bit numbers, and there are a number of points to note. The answer space, QUOT, must be cleared beforehand, as the result shifted in from the least significant end, is unlikely to reach the most significant end. The subroutines called should be self-explanatory, although not all are included in the text. For example, SHR4 shifts a 4 nibble number one place to the right (towards the least significant end) thus halving the number. If the microcontroller has a shift left instruction it may be more efficient to use this rather than the ADD routine where numbers have to be doubled. At the least it would save having to specify one of the two operand addresses.

Arithmetic to other bases

Addition

It is quite common to have to perform arithmetic to the base ten, and not infrequent to have to use some other base, as for instance when working with time, or non-metric weights and measures. Many of the microcontrollers available have instructions which simplify base ten operations, and some have instructions which make it easy to use any required base. It is not difficult, though, to perform these operations even if the microcontroller only has straightforward binary operations available. To understand the process for decimal or BCD, consider adding two nibbles, or rather digits as they are in this situation. The input values can range from 0 to 9, and there may be a carry from a previous stage. The maximum sum can therefore be 19. The two numbers and carry are added together as if they were binary numbers. The result can lie in one of three ranges – under 10, 10 to 15, and 16 or over. If the result is 16 or over, the binary addition will generate a carry, and the true answer is obtained by adding 6 to the result of the binary sum. 6 is the difference between 10 and 16, the effective base of 4 bit binary arithmetic. If the result is between 10 and 15, no carry is generated from the binary sum, but if 6 is added to get the true result, carry will be generated. If the result of the binary sum is below 10, no adjustment is necessary, and the carry value, 0, is correct. So by examining the carry bit and adjusting by 6 it is possible to perform the decimal addition. Here is an example of code to do this. The first operand is in the accumulator on entry, and the second operand is in memory, with the pointer set to point to it.

```
DECADD: ACSC                   ;Add 2nd operand plus carry to A
                               ;skip if carry set as a result
        JMP     DCA1           ;jump if result < 16
        AISC    6              ;Add 6 to A. (There can be no carry from this)
        ST                     ;store result
        RT
DCA1    ST                     ;store result, it may be right
        AISC    6              ;add 6, skip if carry
        RT                     ;No carry, the result stored is right
        ST                     ;store adjusted result
        SC                     ;set carry
        RT
```

The result overwrites the second operand, and the carry bit is valid for summing the next most significant digit. Note that while the ACSC instruction sets the carry bit, the AISC instruction does not, hence the need to set the carry bit before the final return. In other instruction sets it will be different, and it might be necessary to set carry before the first return statement (on the line before DCA1). The principles used here can obviously be extended to counting to any base less than 16, and a routine such as this can be called from an addition routine like ADD4 to perform multi-digit arithmetic.

In microcontrollers where there are special instructions for dealing with decimal arithmetic it is much more straightforward. In some there is a DAA instruction (decimal adjust on addition). Following the addition of two numbers, the DAA instruction is obeyed, and if the result of the addition is greater than 9, or if carry has been set as a result of the addition (the two cases tested for in the routine above), 6 is added to the result to give the correct decimal answer. This obviously only applies to decimal arithmetic, and counting to other bases will still have to be treated as above with the appropriate number used for the adjustment. In some microcontrollers there is a decimal add instruction which performs the addition and decimal adjustment all in one go. Again, this is of no use when working to other bases, but there is a useful instruction available in some microcontrollers, called 'N Adjust' which will add in a carry bit and adjust the result to any base. This certainly simplifies the code when adding 1 on to a time, say, where any carry propagates through the number, and some digits are to base ten and some to base 6.

Subtraction

Subtraction is treated in a similar way. If there is no special provision for decimal arithmetic, a simple group of instructions serves the purpose. It is in fact simpler than addition, and can be performed in three or four instructions with the most basic of microcontrollers. If the subtrahend is in the accumulator, and the minuend address is in the memory pointer, and the carry bit is valid as 'Not Borrow', the code consists of:

```
        COMA                ;complement Acc
        ADCS                ;Acc + memory + carry, skip if carry
        ADX     10          ;decimal adjust if no carry
        NOP
```

The NOP has to be included here as the ADX 10 instruction skips if there is a carry from the addition. The ADX 10 does not however affect the carry flag, which will have been cleared if the ADX 10 is being obeyed, indicating the borrow condition.

Some microcontrollers have a decimal adjust for subtraction (DAS) instruction. Like the DAA instruction, the DAS is used after the binary subtraction and in one instruction performs the adjustment if it is required. Other microcontrollers have the complete decimal subtraction operation available in one instruction. Subtraction to other bases will usually have to be carried out 'long hand', using code similar to that shown above with a different adjustment value. In some microcontrollers there is the 'N adjust' instruction for subtraction, which simplifies carry propagation by subtracting the carry and adjusting the result to the specified base if necessary.

Multiplication

Multiplication to base 10 involves an algorithm which is slightly different from the binary multiplication example given above. Multiplication involves repeated addition, and in decimal multiplication that has to be decimal addition. Also, in the binary multiplication the multiplicand was doubled at each stage, as the weight of each successive bit in the operation is twice that of the previous one. This is not true, though, in decimal multiplication. If a bit by bit approach is used, at every fourth bit the multiplicand has to be multiplied by 10/8, or 1.25. This can be achieved by saving the value from two bits previous and adding that on, but the method is rather clumsy. Instead, the usual algorithm used works on a digit by digit approach. For each digit of the multiplier, the multiplicand is added to the result a number of times, equal to the value of the multiplier digit. The multiplicand is then multiplied by ten, or shifted up 4 bits (which has the same result when working with BCD digits), and the process repeated for the next digit.

The example below, which is written for a Toshiba processor, takes this approach, but saves on execution time at the expense of RAM space by changing the value of the pointer instead of shifting the numbers. The instruction set for this family of processors has a number of unique features. There is one memory pointer for indirect addressing, the HL register pair. This pair of registers can be loaded with an instruction LD HL,x, and exchanged with a memory value with the instruction XCH HL,x, where x is the memory address (restricted to 8 bits, the least significant two being zero). The L register is loaded from the data at address x, and the H register is loaded from the data at address x+1. Another feature is the branch facilities. All branches are conditional on a status flag being set. Some instructions, e.g. LD A,@HL (load accumulator from memory at address HL), always set the status flag, while with other instructions the value of the status flag is dependent on the outcome of the instruction, e.g. after the ADD A,@HL (add the contents of memory at HL to the accumulator), the status flag is set if there is no carry from the addition. The memory for this routine is arranged thus:

	0	1	2	3	4	5	6	7	8	9	A	B	C	D	E	F
0	PTR0L	PTR0H	COUNT	-	PTR1L	PTR1H	-	-	PTR2L	PTR2H	-	-	TEMPL	TEMPH		
1	0	0	0	MPC0	MPC1	MPC2	MPC3	0	0	0	0	-	MLT0	MLT1	MLT2	MLT3
2	RSL0	RSL1	RSL2	RSL3	RSL4	RSL5	RSL6	RSL7	-	-	-	-	-	-	-	-

Where PTR0L and H are the low and high nibbles of the pointer to the multiplier, MLT0−3, PTR1L and H are the pointer to the multiplicand, MPC0−3, and PTR2L and H are the pointer to the result, RSL0−7. COUNT is the digit counter, and TEMPL and H are used for for a temporary pointer in the add routine. By positioning the multiplicand between zero digits as shown, the weight of each digit can be altered just by changing the pointer when calling the add routine. The multiplier is positioned at the top of a 16 nibble page so that incrementing its pointer will generate carry to indicate that the routine is finished. Prior to calling the routine the result must be set to zero and the pointers have the following initial values:

PTR0 = 1C, PTR1 = 13, PTR2 = 20.

The multiplication routine is coded thus:

```
;Decimal Multiplication

DMUL:   LD      HL,PTR0      ;point to multiplier digit
        LD      A,@HL        ;get multiplier digit
        TESTP   ZF           ;test zero flag
        BSS     DML2         ;branch if digit = 0
        DEC     @HL          ;decrement digit
        CALL    DA88         ;add multiplicand to result
        BSS     DML1         ;status = 0 after subroutine
DML1:   BSS     DMUL         ;status = 1 after BSS
DML2:   ADD     PTR1,#F      ;subtract 1 from PTR1
        ADD     PTR0,#1      ;add 1 to PTR0
        BSS     DMUL         ;branch if no carry
        RET                  ;routine finished if carry set
```

The routine DA88 called above performs decimal addition on two 8 digit numbers. As there is only one data pointer, HL, the routine uses a temporary store TEMP for the pointer not in use, and swaps the two pointers between HL and TEMP. The initial pointer values thus remain unaltered in PTR1 and PTR2.

```
;Decimal Addition of two 8 digit numbers

DA88:   LD      HL,PTR2      ;load pointer to 2nd operand
        XCH     HL,TEMP      ;put pointer in TEMP
        LD      HL,PTR1      ;pointer to 1st operand
        ST      #8,COUNT     ;set count value to 16 - digit count
        TEST    CF           ;clear carry flag
DA881:  LD      A,@HL        ;get 1st operand
        XCH     HL,TEMP      ;address 2nd operand
        ADDC    A,@HL        ;add operands + carry
        BSS     DA882        ;branch if no carry
        ADD     A,#6         ;decimal adjust
        BSS     DA883        ;never carries, will always branch
```

```
DA882:  ST      A,@HL       ;store result
        ADD     A,#6        ;decimal adjust
        BSS     DA884       ;branch if no carry
        TESTP   CF          ;set carry flag
DA883:  ST      A,@HL       ;store adjusted result
DA884:  INC     L           ;increment pointer
        XCH     HL,TEMP     ;address 1st operand
        INC     L           ;increment pointer
        ADD     COUNT,#1    ;increment COUNT
        BSS     DA881       ;branch if no carry
        RET                 ;otherwise return, routine complete
```

This routine follows the same lines as the routine DECADD above as far as the decimal adjustment is concerned. Note that with the instruction ADD, x,#n, the status flag is set if there is no carry from the addition, but the carry flag is unaffected.

Division

Decimal division can be carried out in a very similar way, subtracting the divisor from the dividend, counting the number of successful subtractions, and using the remainder as the operand for the next decade. It is useful to check that the divisor is not zero before the routine begins, and it may also help, depending on what instructions are available, to replace the divisor by its 10s complement, so that a decimal add instruction will effectively perform a decimal subtraction.

Logarithmic operations

National Semiconductor, in their data book on microcontrollers, publish an application note describing a technique for multiplying, dividing, and other mathematical operations, based on using logarithms to the base 2. As the method is applicable to any microcontroller it is worthwhile describing here.

The logarithm of a number is the power to which the base must be raised to give the number. To find the exact value of a logarithm involves a lengthy calculation, summing a series and often working in floating point arithmetic. This process involves multiplication and division, and is obviously not a short cut to either. However, it is possible to obtain approximate values for $\log_2$ without any such calculation, by a method which can make multiplication and division much quicker, and can make it possible to calculate other functions, for example roots, by reasonably short algorithms.

The basis of the method is to find the characteristic of the $\log_2$ (the whole number part) by the simple procedure of shifting the number up to find the position of the most significant 1. The mantissa (the fractional part) can then be estimated from the remaining number. At the simplest level this estimation consists of a linear approximation, using the remaining number as it is. This leads to quite large errors, but might be useful in some systems. For numbers which are integer powers of 2, that is 1, 2, 4, 8, 16, etc, the $\log_2$ which is generated is exact, and the maximum error lies halfway between each pair of these numbers. For multiplying 8 bit numbers to give a 16 bit product, the maximum error is at 192 × 192, where the error is 11.11%, whereas, over the whole range the average error is about 3.75%.

A routine for calculating $\log_2$ for an 8 bit number might look like:

```
ULG2:   LBI     7        ;Count 7 to reg B
        REC              ;reset carry
ULG21:  LAM              ;load A from memory (lsn)
        ROTL             ;Shift left 1 bit, msb to carry
        LMAIY            ;store shifted data and address msn
        LAM              ;load A with msn
        ROTL             ;Shift left, msb to carry
        LMADY            ;store shifted data and address lsn
        TC               ;test carry
        BR      ULG22    ;Branch if carry set
        DB               ;decrement count
        BR      ULG21    ;Branch if no borrow
        SEC              ;Reaching this point
;indicates that the original number was zero, so return with carry set.
        RTN
ULG22:  REC              ;Characteristic now in B reg
                         ;Mantissa replaces original number
        RTN              ;return with carry clear with good
                         ;                          result
```

The memory pointer is set up before calling the routine to point to the less significant nibble of the 8 bit number whose $\log_2$ is to be found.

The more significant nibble is in the adjacent memory location addressed by incrementing the Y register. On exit, if carry is clear indicating a successful conversion, the number has been replaced by the mantissa of the $\log_2$, and the B register holds the characteristic. If carry is set on return from the routine it indicates that the original number was zero, and hence no log can be found. To perform the Antilog function the number has to be shifted right until the most significant 1 is in the correct position as indicated by the characteristic. At the same time the 1 that was shifted out into the carry bit must be restored. A full routine to take two 8 bit numbers and generate the 16 bit product could be coded thus:

```
LMUL:   CALL    ULG2    ;get log of 1st no.
        TC              ;Good result?
        BR      LMUZ    ;branch if not
        IY              ;address next location
        XMB             ;store characteristic
        XSPXY           ;address 2nd operand
        CALL    ULG2    ;get log
        TC
        BR      LMER    ;Branch if zero
        IY
        XMB             ;Store characteristic
        DY
        DY              ;address lsn
        XSPXY           ;1st operand
        DY
        DY              ;lsn
        CALL    ADD3    ;Add 2 3-nibble numbers
                        ;on exit address pointer
                        ;points to msn of result
        LBM             ;get characteristic to B
        COMB            ;complement B to give count for right
                        ;                                 shift
        DY              ;address mantissa msn
        SEC             ;set carry
LMUL1:  LAM             ;get msn
        ROTR            ;rotate right
        LMADY           ;store shifted data and address next
                        ;                                nibble
```

```
        LAM
        ROTR
        LMADY
        LAM
        ROTR
        LMADY
        LAM
        ROTR
        LMAIY
        IY
        IY                      ;address msn again
        REC                     ;reset carry
        DB                      ;decrement count
        BR      LMUL1           ;branch if no borrow
LMUZ:   RTN                     ;result overwrites 1st operand
LMER:   XSPXY                   ;address 1st operand
        DY
        LAI     0               ;zero result
        LMADY
        LMADY
        RTN
```

Note that the storage in memory is arranged to be convenient for this routine, with two nibbles of zero adjacent to the first operand to take the two extra nibbles required by the result. The execution time of such a routine is data dependent, but typically this routine would take about half the time taken by the equivalent routine using the technique of MUL: above. To make it more useful, however, it does need to be more accurate. There is a simple method of improving the accuracy to better than 2% overall for 8 bit numbers, with an average accuracy of better than 0.5%. This method involves using a look up table to apply a correction factor to the simple 'linear' $\log_2$ found above. The term 'linear' is used as it makes the assumption that the log function is a straight line between exact powers of 2. In fact it is a curve, and so that value can be improved by adding a small correction factor determined solely by its position relative to the adjacent exact power of 2, in other words by the value of the mantissa. By using the mantissa value, or part of it, as the address for a look up table, and adding on the quantity given, the mantissa can be corrected to any required accuracy. The accuracy indicated above, better than 2%

overall, can be achieved using a look up table with sixteen values. A similar correction is required in the antilog process. For best results the correction table for antilogs should be different from that used to create the log values. The method can be employed with a single look up table but the accuracy suffers. The values under 'Single correction' below give a maximum error of about 2.4%, and an average error of 0.65% for multiplying 2 8-bit numbers. The table below shows the values which give the best results. The correction values are expressed as 8 bit numbers (in hexadecimal). Note that these are different from the values given by National Semiconductor. There are some misprints in the National application note (in the 1989 edition of their Microcontroller Databook), but even allowing for that, the values given here should give slightly better results. (The average error rate increases to nearly 0.7%, and the maximum to about 2.8%, using the National values.)

Table entry no	Log correction	Antilog correction	Single correction
0	03	02	03
1	08	07	08
2	0D	0A	0C
3	11	0D	0F
4	13	10	12
5	15	13	14
6	16	15	16
7	16	16	16
8	15	16	15
9	14	16	15
A	13	15	14
B	10	13	11
C	0E	10	0F
D	0A	0D	0B
E	07	08	07
F	03	03	03

It is not a difficult task to generate these tables using a simple Basic program on a personal computer. The $\log_2$ is given by the formula:

$$\mathrm{Log}_2(x) = \mathrm{Log}(x)/\mathrm{Log}\ (2).$$

Where Log(x) and Log(2) are to the same base. The value of the base is irrelevant, so either $\log_{10}$ or natural logs can be used. If the characteristic of this log is C, then the mantissa M of the 'linear' log is given by:

$$M = (x - 2^C)/2^C.$$

Using these formulae it is possible to generate the true $\log_2$ and the 'linear' log of any number, and compare them. The table of correction values for logs above was established by generating all the values of $\log_2$ and 'linear' log of numbers from 1 to 256, and calculating the average error for mantissa values falling into each of the sixteen bands given by examining the top 4 bits of the mantissa. For the antilog table, calculations were performed for 256 $\log_2$ values in the range 8 to 9. For each value $v = \log_2(x)$ the corresponding value of x was found ($x = 2^v$). The 'linear' log of x was then calculated, and the difference found. These differences were then averaged in blocks of 16 to give 16 average correction values which, when subtracted from a $\log_2$, will give a close approximation to the corresponding 'linear' log.

As an example consider the evaluation of 143 × 76:

> 143 is 8F in hexadecimal, or 1000 1111 in binary. (The space is included in the binary value to make it easier to read.)

This $\log_2$ is calculated as follows:

```
     Start with a Characteristic of 7
L1:  Shift the number of the left, examine carry
     If Carry is 1, go to next stage, L2:
     If carry is 0, decrement characteristic, go back to L1:
L2:  Number remaining is mantissa, use top 4 bits as table address.
     Add table value
```

For 143 this gives a characteristic of 7 and an initial mantissa value of 1E. The correction table indicates that 08 should be added to this to give a final value for $\log_2(143) = 7{,}26$. The comma is used here to indicate that this is not a normal decimal number, but the two parts of the $\log_2$ expressed in hexadecimal. The decimal equivalent is 7.1484.

For 76, or 4C hex, 0100 1100 binary, this gives a characteristic of 6 and an initial mantissa of 30, which changes to 41 when the table correction value of 11 is added.

The product is given by adding the logs:

$$7{,}26 + 6{,}41 = D{,}67$$

D,67 is corrected from the table by subtracting the antilog entry for 6 which is 15, to give D,52. The antilog procedure is then followed. The mantissa, 0101 0010 in binary, now represents a sixteen bit number, obtained by adding 8 0s. The number is then shifted right by (16 – characteristic), and at the first shift a 1 is put into the most significant bit, thereafter 0s are shifted in at the top. This gives a final value of 0010 1010 0100 0000, or 2A40 hex, or 10816 decimal. The exact answer is 10868, so the error here is 0.48%.

Strictly speaking, of course, the mantissa is the fractional part of the logarithm, and the true significance of the binary bits is that the most significant bit represents the value 0.5, the next most significant bit represents 0.25 and so on. Rearranging the formula for the mantissa above,

$$M = (x - 2^C)/2^C$$

gives:

$$x = (M + 1) \times 2^C$$

which is the antilog formula. In the process above, the mantissa was effectively multiplied by 2^{16} by placing it into the 16 bit field. Shifting it right by the difference between 16 and the characteristic adjusted the value so that the overall effect was to multiply by 2^C, and the 1 shifted in at the top with the first shift performs the addition of 1 indicated in the equation above.

Having established the method, it remains to code it. The way this is coded very much depends on what is on offer in the instruction set being used. The code above is written in Hitachi H400 series code, and the instruction set for the H400 includes a 'P' instruction, (for pattern generation) primarily intended for looking up screen segment patterns, e.g. seven segment, from numerical codes. It will however work equally well for this application. Many microcontrollers do have some sort of table look up instructions, but if one is not available, the next best thing is a table jump. With a table jump, the destination of the jump depends on the data, so data dependent code can be executed. Examples of both types of routine follow. The first, in Hitachi code, uses the 'P' instruction, and could be added on to the routine ULG2 above, immediately before the final RTN:

```
        IY                      ;address msn of mantissa
        LAM                     ;Mantissa msn to Acc
        IY                      ;address characteristic store
        XMB                     ;store characteristic
        LBI     TBLAD1          ;Table address bits 4 - 7 to B
        DY
        DY                      ;Address Mantissa lsn
        P       TBLAD2          ;Table address bits 8 - 11
                                ;correction table data to A & B
        AMC                     ;add lsn mantissa + correction
        LMAIY                   ;store new value, address msn
        LAB                     ;correction msn from B to A
        AMC                     ;add msn mantissa + correction
        LMA                     ;store corrected mantissa
        BR      ULG23           ;branch if carry set
        IY                      ;address characteristic
        LBM                     ;restore to B
        RTN
ULG23:  IY                      ;address characteristic
        LBM                     ;Characteristic to B
        IB                      ;increment characteristic
        RTN
```

The table has to lie on a 16 address boundary for the top four bits of the uncorrected mantissa to be used as the bottom four bits of the address.

A similar section of code is required in the antilog procedure, subtracting the correction before performing the $(M + 1) \times 2^C$ operation.

If the microcontroller instruction set does not include proper table look up instructions, but has a table jump instruction, the code will be a bit longer. The section of code below illustrates the type of procedure that would have to be followed in such a microcontroller. The number resides in addresses 0 and 1 of the current 16 nibble page, addressed in the form '0P' and '1P'. The characteristic is stored in address 2.

```
LOG2:   MOV    #7,2P     ;preset characteristic
LG21:   ASL    0P        ;shift number up
        ROL    1P        ;top bit to carry
        BCS    LG22      ;branch if carry set
        DEC    2P        ;decrement characteristic
        BCC    LG21      ;branch if carry clear (no borrow)
        JMP    LGERR     ;number is zero
LG22:   JMP    @1P       ;table jump on mantissa msn
        JMP    COR0      ;mantissa msn = 0
        JMP    COR1      ;mantissa msn = 1
        JMP    COR2
        .
        .
        etc
COR0:   ADD    #3,0P     ;add correction factor of 3 to
                         ;                    mantissa
CR00:   BCC    CR01      ;branch if no carry
        INC    1P        ;propagate carry
CR01:   RET              ;corrected log in 0, 1, 2
COR1:   ADD    #8,0P
        JMP    CR00
COR2:   ADD    #D,0P
        JMP    CR00
etc.
```

As there are repetitions in the correction table it is not necessary to include all 16 addition sections. The antilog is found in a similar way, subtracting the correction factors then shifting to find the antilog.

The use of logarithmic techniques can considerably reduce the time taken by calculations. Usually the code required occupies more memory space, and the results are not exact, but there are many applications where speed is the most critical factor, and these disadvantages can be tolerated. As an example of the speed gains to be made, consider finding a square root. In their software application notes, Hitachi give a square root routine which will find the square root of a sixteen bit number in about 1500 cycles. Using the logarithmic method this could be reduced to a typical time of about 350 cycles, the exact time being data dependent.

This logarithm method is obviously also applicable to division, where it is very useful, particularly when the numbers involved are large, but the result is known to lie, say, within the 8 bit number range. It gives an almost floating point 'feel' to the calculations, and executes in considerably shorter time than the exact method. It always has to be borne in mind that this method is not exact, and can only be used where the errors introduced can be accommodated.

The usual reason for employing the logarithmic technique is to speed up calculations. However, there will be situations where even this method is not fast enough, and an alternative will have to be found. The quickest way is usually to employ a look up table. The results are precalculated for all possible input values and stored in a table. The input value is then used as an address to access the appropriate result. The method is quick, but the look up table can be long, occupying more space than can be afforded. There is often a compromise solution, which is to use both methods. The calculations can be performed over a certain range of input values when there is time for them to be completed, and the look up table only then has to be used for a limited range of values, thus limiting the space required by the look up table. Take for example a tachometer type of application, where the rate of input pulses has to be calculated from a measured pulse period. The rate has to be calculated within one pulse period. At low speeds, when the pulse period is long, there is plenty of time for the calculation to be performed. At high speeds, with a short pulse period, there is not enough time to perform the calculation, and so a look up table method must be used. There will be a point somewhere in the mid range where the calculation can just be performed, which can be defined as the switch over rate. The incoming period measurement is compared with the switch over rate, and if it is less the calculation is performed, but if it is greater, then the look up table will be used. In this way the look up table is kept to the minimum size necessary. Care must be taken to ensure that the correct switch over rate is chosen. Allowance must be made for calculation times being extended by any interrupts which the system has to deal with, and the worst case scenario must be defined, with the processor clock at the extreme where the calculation would be at its slowest. In this example there is an obvious alternative, which may be applicable in some cases, especially if an average rate is required. The rate can be calculated over several input pulses if a true instantaneous rate is not required. The number of input pulses could be fixed to ensure that the calculation can always be performed in the available time, or a method using a variable number of input pulses could be used. The calculation proceeds, while input pulses are counted in by an interrupt routine. When the calculation is finished, the data for the next calculation is ready waiting, giving the number of pulses received, and the total time for them. The advantage of this method is that the results are available as fast as the system can calculate them. With the fixed number of pulses method, the results may be unacceptably slow at low rates, just to accommodate the extreme upper range.

10 The system development environment

This chapter examines the process of developing a system or product from its original inception to the end product reaching its intended user. The stages that are usually passed through in the process are explored, and the different types of equipment available to enable each stage to be completed in the most cost effective manner are reviewed.

Stages of development

Very often a new product is a development from a previous one, taking advantage of advances in technology to improve the product in some way. Reducing size and cost, adding new features, or improving the operator interface are typical examples of such improvements. Sometimes, though, a new product is completely new. It may have been an idea that has been waiting for technology to advance sufficiently to make it feasible, or it may be the result of some new demand from possible users, or it may just be the implementation of some bright new idea. Whatever the source, if it is something completely new, it may be necessary to proceed in a fairly cautious manner to ensure that the final product will be able to do what it is supposed to do, and that it is possible to manufacture it within reasonable cost limits. Even if the new product is a development from an existing model, it is likely to have some novel aspects which need to be investigated in some depth before the main development begins. In order to start the development process, a feasibility study is often undertaken. Usually this is just a paper study, but it may be necessary in the course of studying the feasibility of some new idea, to build some sort of concept demonstrator, a model which will demonstrate that certain critical aspects of the system, ones which have not been performed before, or are to be performed in a novel way, do indeed work as was anticipated. This concept demonstrator thus becomes the first stage of a development process. It may have to show the feasibility of some completely new approach, and is quite likely to be the most technically demanding aspect of the development. Yet funds will at this stage be severely limited, as financial backers of any development need to be assured of the technical feasibility of the project before any substantial sums of money are invested. The purpose of this stage of the development is to ensure that the project is technically feasible, and to satisfy all concerned that it is a worthwhile project in which to invest.

In planning a development project, this initial study phase should yield some valuable clues about the costs and timescales needed for the development, and indicate how accurate were the initial estimates. Clues can sometimes also be gained in these early stages as to what level of sophistication is going to be required from the software development system.

Having established that the product or system is technically feasible, and financially viable, the development proper can begin. At this stage there should be a clear idea of what the development hopes to achieve in terms of an end product, and there should be a specification which will bear a close resemblance to the final version. Specifications always do develop with the product, but at the early stages it is sometimes very difficult to know exactly what the specification should be. Some aspects of the specification may be difficult to define, for one reason or another, and to help define the specification it may be necessary to build some test beds to try out certain aspects of the design. Perhaps several alternative approaches to a section of the system need to be tried out to see which is the best. This is particularly true when defining an operator interface. It may be necessary to have test bed models where trials can be held with prospective operators to establish which version is most widely acceptable, and the easiest to use.

After the test bed models, which like the concept demonstrator will probably not model the system in its entirety, comes the first complete system, or prototype version. How closely this mirrors the final form of the system depends on a number of factors, and perhaps the term 'development model' would be better than 'prototype' to describe this stage. It has to include all the functions that are required of the end product, but not perhaps in their final form. If there is to be a display, for instance, it may have to be custom made. The final form of the display may not be known until the development model is in use, and to hold up development while a special display is made would be unnecessarily costly and time consuming. Instead, a display will be made up from discrete elements, or a similar display modified to reflect the functional requirements of the new system. This development model is the version on which most of the work will be carried out, and there may have to be more than one development model to allow all members of the development team as much access to the models as they require in order to work efficiently. The development model will also probably be the main medium for approvals, both of the specification, and of the functional performance of the system.

Once the performance of the development model has been approved, all parties involved having satisfied themselves that the system has been specified correctly, and implemented according to the specification, the next stage is to build preproduction models. It may be that these models are of such advanced form that they do in fact form the first part of a production run. This will be particularly likely to happen if the only major difference is that the microcontroller in the preproduction model is user programmed rather than mask programmed. At this stage all tooling costs will have been committed, and it is probably only a matter of time before the final production versions appear.

For each of these stages, from concept demonstrator to final production version, the features required of the development environment will be different. At the early stages, before it is known whether or not the project is even going to proceed as far as a finished product, it is important to minimise costs, while at the same time ensuring that sufficient technical support is available. The results of a feasibility study or concept demonstration will only be valid if all the capabilities and limitations of the target controller are known and understood. Later on in the project, when development is in full swing, it may be desirable to employ a much more sophisticated development environment, one which will give the engineering staff all the tools they need to work efficiently and accurately. At this stage the major costs involved are probably in the man hours that have to be invested to achieve the project goals, and equipment costs are secondary. To economise on equipment costs here would be a false economy, as it would lead to increases in costs of the engineering manpower, and hence, of the project as a whole.

Program development equipment

All manufacturers of 4 bit microcontrollers provide some form of support equipment for designers to use when developing systems using their parts. Some independent suppliers of development equipment also provide products to support 4 bit microcontrollers. In general the equipment is designed to operate in conjunction with a personal computer, and comes in a number of distinct sections. The heading to this section, 'Program development equipment', reflects the fact that the primary aim of most of the equipment available is that of developing the software for the microcontroller. It does, however, have to operate in a hardware environment, and much of the development of a system may well be hardware development, so the two aspects of the development are closely enmeshed. The equipment available usually allows the user to build the hardware and fit on to it, instead of the actual microcontroller, a piece of equipment which will emulate the actions of the microcontroller, allowing code to be developed and tried out in the target hardware environment.

The first stage in this process is the 'cross assembler', which allows code to be written on the personal computer, usually in the form of an assembly language program. From that program the cross assembler produces a file containing the binary code required by the microcontroller. This is then transferred to the emulation system via a communications link, or sometimes by programming an EPROM, or a programmable version of the microcontroller. Some manufacturers also have simulation programs available. These allow the program to be tried out purely on the personal computer, as a software model, without having to use any hardware at all. They cannot operate in real time, nor can they give a realistic view of the I/O and display drive that may eventually be required, but they can give a degree of confidence in a program, and are useful in checking out parts of the program which perform calculations, and where it is helpful to be able to view intermediate results and not just what may eventually appear on an output.

For development hardware, most manufacturers provide equipment at two or three levels of complexity (and hence cost). At the simplest level there is what is usually described as a prototyping system, a circuit board with a socketed programmable version of the microcontroller, and connectors for I/O and display. Some other generally applicable circuitry may also be supplied, and there will usually be an area of the board where the user's own peripheral circuits can be added on.

The layout of the board is usually arranged to make it easy to attach logic analysis equipment for fault finding, or monitor the behaviour of the system in some other way. At the next stage in complexity is the evaluation board, or kit. The complexity of this piece of equipment varies from manufacturer to manufacturer, but generally speaking this has a section of circuit which emulates the target microcontroller, and allows the user to change the program memory contents with a greater or lesser degree of convenience. The most convenient can be connected to the personal computer, and programs transferred over the communications link. Almost all will also allow the use of EPROMs for storing programs. Sometimes the evaluation board, as well as being available for use as a stand alone unit, is also employed as a 'personality module' for the top of the range development equipment, the in circuit emulator. Typically, a manufacturer will produce one piece of emulation equipment for a family of devices, and for each device in the family a different personality module provides the functions unique to that particular family member. This is true of some of the more sophisticated evaluation kits as well as in circuit emulators. The in circuit emulator, being at the most complex and expensive end of the development equipment range, should provide everything that is needed by development engineers to enable them to work quickly and efficiently. The emulator communicates with the personal computer, from which it collects programs, and through which the user controls the operation of the emulator. The good emulator will allow the development engineer to discover exactly what is going on inside the microcontroller, preferably in real time.

It is probably worthwhile examining each of these items in more detail, to see what choices there are when building up a system development environment.

Assemblers

Chapter 6 contained some observations about assemblers and what facilities to look out for when choosing one. Usually the architecture and integrated peripherals of a particular microcontroller will carry more weight in the decision process than the facilities available from that particular manufacturer's assembler. Nevertheless, it is important to be aware of what is offered in the assembler, and to assess whether or not it is going to be adequate. It is possible to obtain cross assemblers from independent software houses, but in general these tend not to have all the facilities that are available from the specific assembler provided by the manufacturer of a microcontroller. The situation with 4 bit devices is quite different from the 8 bit scene, where there are a small number of basic architectures, with a good selection of second sources, and manufacturers competing to provide the best integrated peripherals with the same basic core processor. In the 8 bit world there are many independent, general purpose development systems available, and a range of independent, processor specific, cross assemblers to run on a variety of host processors. Because of the great variety of 4 bit microcontrollers available, and the tendency for these to be single sourced, there is a much smaller market for any individual cross assembler. As a result the independent assemblers that are available are not specific at all. The user has to 'customise' the assembler by providing it with a list of instruction mnemonics and corresponding binary codes. Consequently, the resulting assembler tends not to be able to provide the sorts of facilities that could be expected from an assembler written with a specific microcontroller, or family of microcontrollers, in mind. In addition to the facilities discussed in Chapter 6, it is worthwhile checking that the assembler includes all the output listing, cross referencing and error handling facilities that are required. It is often difficult to assess the effectiveness of the error handling functions of an assembler until it is used, but an examination of the list of possible error messages will give some clues as to how 'user friendly' the error handling is. Conditional assembly is another facility often provided, and if a project is aiming to produce a number of variants of a basic product it may well be a useful attribute to the assembler. If the facility is likely to be needed it should be examined to see how easy or difficult it is to use.

Simulators

Some manufacturers provide simulators for their products, which allow the behaviour of a particular microcontroller to be simulated on a personal computer. With this method, programs can be checked out without having to build a hardware system for the program to run on. This is obviously a useful way of learning about the capabilities of a microcontroller, and gaining confidence in its capabilities. A simulator will not be able to operate in real time, but it should be able to give some indication of the time taken by a routine or section of code. Whether or not a simulator can cope with a situation where a system is interrupt driven, and can simulate the effects of timer interrupts, and time critical routines, needs to be examined carefully if this is a significant requirement of the system. Using a simulator will also allow more rigorous checking of a section of the software than can sometimes be achieved in the target hardware system. For instance, in a simulation model, test patterns can be generated so that a particular input can be tested at all possible values, or combinations of input conditions can be generated that might be difficult to produce in the full system, but need to be tested to ensure correct operation of the controlling program. Testing the system software, or part of it, using a simulator in this way may in some cases be essential in order to establish that the software is of a high enough quality for the end product. This is particularly true where the system has a critical role to play in a market sector such as finance or security.

Prototyping systems

Prototyping systems usually provide the lowest cost entry point into the development equipment arena. For a few hundred pounds a manufacturer will provide a board on which the facilities of a particular processor can be explored. There will be very little provided in terms of diagnostic aids, and an external power supply will have to be provided to run the board. Such a set up can be useful for small projects, and small concept demonstrators, where the software requirements are not too complex, and can be made to operate correctly without the sophisticated fault finding aids of the more complex development tools. Software has a habit of behaving in ways which are not expected by the originator, and sometimes the reasons for this are quite obscure, and difficult to locate. This is particularly true of real time systems, where apparently random errors can arise, sometimes at quite rare intervals, when the system depends for its correct operation on a certain timing pattern. This usually means the time relationship between a pair of inputs, where a fault can arise if one input occurs within a certain time of the other. Depending on the brevity of the critical time, and the frequency of the events involved, the fault can be very infrequent in its occurrence, but is almost guaranteed to occur during an important demonstration of the system. It is this type of fault that can be so elusive when working with a simple prototyping system, and so it is perhaps better to limit their use to the other end of the development cycle, where it is often required to take a model off the large development system, tied as it is to the mains power supply, and probably also to the personal computer, in order to operate. The prototyping board will often permit a small version of the system to be made up quite quickly without incurring large tooling costs. This can be very useful where the target product is a battery operated portable device, and some feel for how it will eventually operate is required before all the aspects of the design are complete. The prototype system can operate with code that has been tried and tested on the more sophisticated development system, and if any faults do arise, they can be investigated on the larger system, and hopefully tracked down much more quickly and easily.

Evaluation kits

For some manufacturers this is the primary means of program development, whereas for others the evaluation kit is not much more sophisticated than the prototyping board described above. There is quite a wide variety across the range of evaluation kits as to what facilities can be provided. The term 'evaluation kit' does not really have a hard and fast meaning throughout the range of manufacturers in the 4 bit microcontroller business. However it is worthwhile describing a typical evaluation kit. The kit will consist of a printed circuit board, often a large one, populated with the circuitry to perform the functions of the microcontroller, and allow access to internal registers and memory. There will usually be a choice of program source, the kit operating either from a personal computer, and receiving the program downloaded from this over the communications link, or operating from a program held in EPROM mounted in sockets on the evaluation kit itself. There will generally be connectors for attaching the circuitry which is external to the microcontroller, such as display, keyboard, serial links, and any other I/O, while some circuitry normally external to the microcontroller, such as the oscillator components, will be included on the evaluation kit. Power supplies will usually have to be provided, and the evaluation kit will generally consume considerably more power than the microcontroller it represents. When the kit is under the control of a personal computer there will usually be some diagnostic aids available, such as the ability to set breakpoints, single step through the program, and examine the state of internal registers and memory. The number of such facilities, and their sophistication, will vary quite widely from kit to kit, but they will not, in general, be as advanced as the facilities to be expected on an in circuit emulator. An evaluation kit will usually provide sufficient facilities for the development of quite advanced systems, but usually lacks the real time diagnostic tools that are found in emulators, and it can be quite frustrating for the software engineer trying to find out exactly what is happening, especially in a multi-interrupt situation.

In circuit emulators

The in circuit emulator represents the top level of development aid available to assist the system developer. The emulator usually connects to a personal computer via a communications link, and has a 'probe', often in the form of a plug which will fit into the microcontroller socket on the target system. The power supply is internal, and the whole unit is cased, with a front panel giving some operator controls, selecting different modes of operation. There is usually a degree of processing capability within the emulator, and the personal computer will act more as a terminal to supply commands to the emulator, than as an intelligent controller. With the emulator will be supplied a simple communications program to run on the personal computer, which will allow the operator to give all the commands to the emulator, and allow the emulator access to files on disk in the personal computer. The in circuit emulator will usually cost a few thousand pounds to purchase outright, but in many cases it will be possible to hire one for the duration of a project. Many manufacturers, or their agents or franchised distributors, have emulators available, and if the project promises to put a substantial amount of business in their direction, then they may be persuaded to loan development equipment free of charge. With the prospect of sales of a large number of mask programmed microcontrollers, many manufacturers will be persuaded to make concessions to get a potential customer to choose their product. If a development system is sitting idle, most sales departments would rather lend it out than see business go to their competitors.

The facilities provided by an in circuit emulator should include such things as a variety of different types of break point, the ability to trace in real time exactly what instructions have been obeyed, the ability to step through the program instruction by instruction, and the facility for 'unassembling', or decoding machine code back into mnemonics. It is probably worthwhile examining the types of facility to be expected in an in circuit emulator in a bit more detail.

Break points are points in the program where execution is halted to allow the operator to examine the state of internal registers, RAM, etc, and determine exactly what is happening in the machine. Simple break points are just specified by giving the address of the instruction where processing is to stop, and the emulator will stop every time it reaches that instruction. This is very limiting, though, and most systems provide a much more sophisticated system of conditional break points. The first condition usually applied is a count condition. The operator can specify that the operation should be suspended after obeying a certain instruction a given number of times. This is very useful for code which operates in a loop, and could well be used, for instance, to trap an error where a loop was being obeyed too many times.

Another useful break point condition is on memory access, usually RAM. In this case it should be possible to cause the processing to halt if the RAM is accessed at a given address, sometimes being able to specify whether on writing, reading, or on either operation. An added feature of this type of condition is to allow the RAM data to be included in the decision whether or not to break. A similar scheme with I/O ports is sometimes available, and on some systems there is provision for external signals to be monitored via a set of probes, and then the state of these external signals can be used to effect the break point decision. Ideally, it should be possible to combine all the break point conditions using the usual logical operators (AND, OR, NOT), both for simultaneous operation, or sequential operation. When simultaneous conditions are set up, the break only occurs when all the conditions specified are satisfied at the same time, but with sequential conditions it is possible to specify one set of conditions that must be satisfied before a second set is examined. The first set of conditions then 'arms' the machine to act on the second set. This 'arming' type of condition allows, for instance, a break point to be set in a subroutine, which will only be effective if the routine has been called by a specific 'call' instruction. Most systems allow more than one break point to be active at a time, but there is often a limit of, say, ten active break points. This is usually enough for most operations, but if the number of break points is limited to only one or two, it could cause inconvenience for the user of the system.

Trace facilities enable the user to specify in what way a record of instructions obeyed is created, and to decide what 'historic' data is required to be recorded. Important parameters are those which 'trigger' the trace, that is set it going, those which specify the range of the trace, or which parts of the code should be traced, and those specifying the scope of the trace, indicating what information should be recorded. Trace facilities tend to record vast quantities of data in a very short time if the operator is not selective enough when setting up the trace, and as a result, much time can be wasted ploughing through the trace records looking for a particular event which indicates an error. The best trace facilities allow the user to cut the trace information down to a minimum, so that only the data relevant to the problem being investigated is recorded. It is usually better to record too little information and have to rerun a test, rather than record too much information and spend a long time extracting the relevant piece. Trigger conditions for a trace facility need to be very similar to the breakpoint conditions described above, though should be independent of the breakpoint system. In that way a test can be run where data is collected from the system from a point of interest, and a break point can be put in to stop the system after enough data has been collected. The range of a trace determines which instructions are traced. In some systems there are maps with a one to one relationship between the map and the codespace, so for each address it is possible to specify whether or not the instruction should be included in the trace. This is obviously a very powerful facility, but it can be tedious to set up. It is usually possible to specify blocks of code which should be traced, so that instructions obeyed within a certain range of memory addresses are traced. Sometimes it is possible to specify that certain types of instruction should be traced, for example it may be desirable to trace all the 'jumps' obeyed in a range of code. The scope of the trace, that is, what is recorded in the trace record, is often fixed, and the user has no control over what data is recorded. This leads either to there being too much data, and hence wasting time, or there not being enough data, the vital piece being missed out. It is much better if the user has the opportunity to specify which data are stored in the trace operation.

There is a facility available in some systems which comes as a half way house between a break point and a trace. This facility, which should be able to operate in real time, displays on the screen the data at the point specified, just as if it were a break point, but the system continues to operate normally. In this way a trace of what is happening at a given point in the program can be built up. This gives an extra refinement over what can be seen with the trace facility, as the trace data is not usually available for inspection until the test has been completed, and operation of the system halted. Clearly, this display and continue type of break point can have its disadvantages. If the data takes longer to write to the screen than is available between monitoring points the system will not be able to operate in real time, and this may invalidate the tests being performed, but provided a small amount of data is required, it can be a very useful way of having it presented.

A useful feature of some development systems is a time measurement facility. Points are specified in the program in a similar way to defining break points, and the system measures the time taken, during program execution, between the points specified. This is very useful when establishing the percentage of the processor usage which is taken up with processing interrupts, for instance, and for ensuring that the program is capable of performing to a given time requirement. Variations in timing can be measured, and estimates of worst case timings made. This allows the designer to check that sufficient margin has been allowed to ensure that the microcontroller will be able to cope under all circumstances. There may be limitations on the length of the times that can be measured, so if it is necessary to perform time tests over a long period it is wise to check that the facility will be able to do what is required.

The interface with the user is another important aspect of a development system. Often the commands to the system are somewhat cryptic, and if there are many of them, the operator spends a lot of time, especially when new to the system, searching the instruction manual. Sometimes facilities which could be usefully employed are never discovered because of the complexity of the command set. To overcome this, manufacturers are now introducing systems which are more user friendly, and make use of the windows environment on the personal computer. Commands are selected from a menu by use of a mouse, and the menu is updated at each stage to present the operator with a relevant set of possibilities. This is certainly quicker to someone new to the system, and ensures that no facilities are unaccessible because of command complexity. To the experienced user the advantages of a windows type interface are not so definite, and with the older types of personal computer this type of interface can prove rather slow. Another aspect of the user interface sometimes provided is an on screen 'Help' facility. The user presses the appropriate function key, and a message appears on the screen explaining the facility being used. It is important that any such help system is 'context sensitive', in other words it must present relevant information. If it does not present the right information for the context in which it is requested, the user may as well look up the instruction manual.

Another feature very useful in development systems is to have a 'symbolic debugging' facility. When a program is written, symbols are used extensively to refer to constants, variables (in RAM), points in the code (labels), subroutines, etc. It is much quicker and easier for the system designer to continue to use these symbols when transferring to work on the development system. The assembler program, as well as producing the machine code for the emulator to run from, produces a symbol list, giving all the symbols used in the program, and their corresponding values. They may also be given types (label, variable, constant, etc), to ensure that they are used in the correct way. If such a facility is available it will certainly make for more efficient use of the development system. It is important, though, that the development system does not place any greater restrictions on the use of symbols than was already imposed by the assembler. There will be a range of restrictions imposed by the assembler. Such things as the number of symbols that can be used, and the number of characters within a symbol, and which characters are allowed to be used in a symbol, will all have their restrictions, as well as having a list of reserved words which cannot be used as symbols (all the mnemonics of the instruction set for example). If the development system imposes further restrictions on top of these, it could prove to be more of a hindrance than a help.

If the program has been written using any higher level constructs, from simple macros to features like 'while' loops, and 'If, Then, Else' type structures, debugging can be very tedious if the designer is forced to operate purely in machine code. If the assembler allows the use of such constructs it should also provide a listing of their expansions into machine code, but this is not always the most helpful way of working. Better still is if the development system can use these higher level constructs, and allow the user to work in code lines rather than machine code instructions. It should be possible to specify break points in terms of original source code, and lines of original source code be displayed for breaks and traces. Together with symbolic working, this facility allows the designer to feel completely at home on the development system, and work to the maximum efficiency in checking the function of the software, and eliminating any bugs.

There is increasing concern nowadays over the quality of software, perhaps because it is being used more and more to take over the functions previously regarded as hardware territory. The concern is that the software may be faulty in some way, and that the fault may not show up for years, but when it does it will prove catastrophic. Concern starts at the point where changing the code becomes very costly, as in mask programmed systems, and continues into systems where the integrity of the system is critical to some degree. It may be in finance, where a software error could again be very costly, or in security, or even in safety critical systems where lives could be put at risk by faulty software. To a certain extent such systems can be built to be fault tolerant, but this is only a partial solution. The subject of safety critical systems is beyond the scope of this book, but there are some facilities offered by some development systems which can be an aid to confidence in software. Such quality assurance facilities may be a requirement of the regulatory authority or licensing board relevant to the equipment under development. Particularly useful facilities are the ability to identify which parts of the code are untested by a set of test routines, and the ability to identify any redundant code, together with a test documentation system, which verifies that the necessary tests have been performed on the system.

The final feature to be mentioned is more a hardware facility than a software one. Most systems need, at some stage in the development, to have the code transferred to an EPROM, whether built into a user programmable version of the microcontroller, or in an industry standard package, or possibly into an OTP version of the microcontroller. EPROMs are often used as a medium for transporting code when specifying the code of a mask programmable device, and systems are frequently transferred to stand alone emulators, evaluation kits, or prototyping systems, for the final stages of testing before committing to a mask programmed device. The facility for programming an EPROM or OTP version of the microcontroller is therefore almost universally required. Many development systems have EPROM programmers built in, so as to simplify the process, and cut out one stage of data transfer. Data is transferred once to the development system, tested on the system, and it is this same code which is then programmed into the EPROM. A separate EPROM programmer can be used, of course, but this involves an extra transfer of the data, an extra opportunity for error, and hence an extra need for checking and testing. The development system with built in programmer can therefore save time and increase confidence in the final product. In addition to this, it may be that the development system will accept the OTP or EPROM version of the microcontroller without an adapter, which would almost certainly be needed if the part were to be programmed on a 'universal' PROM programmer.

It can be seen from the foregoing discussion that there are quite a number of features desirable on a development system, and when making a choice of microcontroller, the facilities provided in the back up systems do have to be considered.

Project control

Overall control of a development project, keeping track of progress, and trying to ensure that the project meets targets in costs and timescales, is never an easy task. Any assistance with project control is to be welcomed, and any system which makes the job easier is to be applauded. One aspect of controlling an engineering project which often involves more time and effort than it should is 'change control', or 'version control'. When putting a product into production it is obviously essential to make sure that (a) the production department knows exactly what to produce, and (b) when any changes, modifications, or improvements are introduced by the development team, the information is transferred quickly and accurately to the production department. It may also be necessary to keep track of the exact make up of each system produced, so that at some future date, by looking up the system serial number, it is possible to say exactly which version of software and exactly which hardware modifications were included in that system. Version control of this nature usually begins with a 'release' of a version from the development department, but there may also be a need for some form of version control internal to the development department, to ensure that team members are always working to the latest standards. As new sections of a system are completed, changes often have to be introduced in previously produced sections to ensure that they work properly together. An individual member of the development team may work on a section of the system, and when satisfied with it, release it to the rest of the team for testing, or inclusion in the current version of the development model. Keeping track of the developments, making sure that all team members are working with the latest version, and ensuring effective communication between team members, relating test results to the appropriate version of the system, and so on, can be a time consuming task for the project leader, who is often intimately involved with the development, and has no time for such monitoring and control. The result can lead to frustration and time wasting by other team members, as faults are corrected and not fed through by one team member into the systems being used by other team members, or new faults are introduced, but not discovered until a late stage.

There are software systems designed to help keep such things under control, and aid the progress monitoring that is needed to ensure that budgets and timescales are met. The application of such systems imposes restrictions and disciplines on the working practices of individuals in a team, and ensures that it is possible to keep track of new design releases, changes made, and tests completed. In the modern development environment, where development team members all have their personal computers linked together in a network of some sort, and where the network may also encompass other departments such as quality control and production control, it will be essential to define a set of working rules to ensure that information flow is correctly controlled, and systems, or sections of systems are adequately tested before being released from one level to the next. It is unlikely that any software system which can introduce this kind of monitoring and control over a project development will come from the microcontroller manufacturer, or from the same source as any of the computing facilities closely associated with the microcontroller and the development of software for it. Most such systems will be designed to allow the user to employ whatever software tools are necessary - editors, assemblers, compilers, linkers, library control, etc. Before investing in such a project control system it will be necessary to ensure that it will indeed be possible to operate it in conjunction with the other development tools which it is planned to use. It may be that the project control system is already in use, and any new software which it is planned to use in connection with a new microcontroller development will have to fit in with, and work alongside, the existing systems. It is possible that there is a quality control system already established that defines working practices, and imposes disciplines on the development of a new product or system, and again the new software tools will have to be able to operate in the context of these existing structures. Whether or not this imposes any peculiar restraints on the way in which the new software has to work depends on the system in use, and it is to be hoped that any set of software tools associated with a particular microcontroller will fit into an existing project control. It is necessary to address this question of software compatibility, and if required, define any changes needed to allow the new tools to be integrated into the overall development environment.

Choices

In any new system or product, the embedded microcontroller is likely to be a key component, having a major impact on the functions available, and the overall cost, not only of the finished article, but of the development as well. Making the right choice of microcontroller is a key decision, and one which has to be made quite early on in the development cycle. As soon as a product specification has begun to be formulated, and the desirable functions and the target costs defined, it is time to begin thinking about how these are going to be achieved, and hence, which microcontroller to employ. Not only does this decision affect the current development, but it also has ongoing implications for future developments. If a range of products is planned, will the microcontroller chosen for the first of the range also be suitable for the last? It may not be appropriate to try and find one microcontroller which will cover all the requirements of a whole range of products, and to do so may put an undue cost burden on the early products of the range, or those members of the range which have the lowest functionality. However, microcontrollers tend to come in families, with various amounts of program memory, data storage, I/O capacity, display drive capability, etc, and there are distinct advantages to being able to use related controllers in related products. A range of microcontrollers from one family should share an instruction set and development support facilities, with only minor modifications, and so much of the capital costs of setting up the development can be shared across the range of products, and it should usually be possible to preserve a central core of the development work, particularly software structures, for use in the whole range. If for some reason it is necessary to use a completely unrelated microcontroller for some members of the product family, it may mean virtually starting again from scratch as far as the software development is concerned. A different microcontroller may mean a completely different organisation of the software, with different interrupt handling techniques, different display drive methods, and a completely different software development environment. This all puts cost on to the development, and this must be quantified in order to reach an informed decision.

Starting out on the development of a new product, or system, is always a point at which to review current practices and procedures. When investing in a new development environment, or extending an existing arrangement to cope with the new requirements, all sorts of questions arise, and it is a good opportunity to examine the way things are being done and to decide whether it is cost effective to change them in any way. Choosing the microcontroller may well determine the nature of the development environment, and that in turn may have an influence on the wider field of working practices, documentation, control, and quality assurance. There may be externally applied standards to which the development has to comply, in order to qualify for application to a certain market sector, and it is important to ensure that the chosen microcontroller fits the bill in all aspects. Sometimes there will be a choice of devices from different manufacturers. It may well be that a family of parts is required for a family of products, and that several manufacturers can supply suitable parts, at very similar prices. In this situation, the development facilities available for each family of microcontrollers, and the design environment that it is possible to build up round a particular manufacturer's products, could be a major deciding factor in the choice of microcontroller.

There are many other factors involved in the choice of a microcontroller, such as the project size, options on purchase, hire, or lease of development equipment, and also the availability and quality of any technical backup from the manufacturer or agent.

Concerning project size, if the project is a small scale development, with only one or two engineers involved, this will mean that resources for the development are probably limited, and the lowest cost route must be employed. With bigger projects, where the software and electronic hardware designers are part of a bigger team involving other disciplines, such as mechanical engineers, materials scientists and so on, the capital cost of the development system may well be less important. More important will be its ability to fit into existing systems, and work alongside whatever communications and project control mechanisms are already being employed. The timescale for the development will also have some bearing on this. If it is of paramount importance that a product be developed in the shortest possible timescales, in order to reach the market in time to capture a significant portion of it, then development cost budgets will be able to be relaxed in the interest of shortening the development time. In this case any techniques for arriving at the end result more quickly will be welcomed, and if extra development systems can be usefully employed to do this, then the ease with which systems can be coupled up to work together could be an important feature.

How the development equipment is to be financed is another complex decision. Clearly, if the equipment can be obtained on loan, free of charge, it is a distinct advantage as far as the accountants are concerned. However, what guarantees are there that the equipment will not be recalled? It is possible that the equipment will suddenly become unavailable at a critical stage of the project. If this is the only equipment of its kind, and is lent out by the distributor from company to company developing products, what are the probabilities of the equipment breaking down at a critical point, and causing expensive delays? If, after returning the equipment, project completed, it is required to use it again for further tests, or more development work, will it still be available? These are all risk factors to take into account, and offset against the advantages of free use of the equipment. The next option is to hire the equipment, and here the problem is to know how long the equipment is required for, and try to calculate if this is the best option. One attraction of the hire option is that usually there will be a piece of back-up equipment available at short notice, should a breakdown occur. On the other hand, if the piece of equipment is fairly rare, the arguments against borrowing may also apply to hiring, but at least the obligations on the hire company should be more compelling than those on a company which loans out its equipment. If it is decided that a piece of equipment is required on a permanent basis, the final choice is whether to lease or purchase outright. This is not a technical decision, and should have no bearing on the technical aspects discussed here, but can be safely left in the hands of the accountants.

The final consideration in the choice of a system is that of technical back-up. When the equipment arrives, its manuals written in impeccable Japanese, will there be anyone there to translate them for you? The majority of the 4 bit microcontrollers on the market today are made by Japanese companies. Most of them have good representation in the West, either in their own plants, technical sales organisations, or design labs. Many manufacturers also make use of franchised distributors, and a condition of the franchise is usually that the distributor provides a certain level of technical expertise and help to customers. It is an unusual development which can go through from inception to completion without the need to draw on the resources of some technical back-up team. The handbooks, data manuals, application notes and software guides are usually able to satisfy the vast majority of technical enquiry, but there is usually some little point of detail which needs to be clarified, or some obscure facet of the operation of the microcontroller which is not well explained, and it is in these circumstances that it is essential to be able to call on someone who knows the answer. No point of doubt, however small it may seem, can be left to chance. It will no doubt come back to haunt the designer in future years if it is not cleared up at the time. It is therefore essential that the necessary technical back-up facilities are available, or valuable time and energy is going to be wasted. If there is no intelligent go-between

between the manufacturer in Japan and the development engineer, there are endless possibilities for delay and confusion, misunderstanding and mistake.

11 A 4 bit microcontroller in use – An example

This chapter is devoted to an example of the use of a 4 bit microcontroller. In this example an OKI 6351 chip is programmed to perform the functions of a watch and stop-watch. The code shown here is part of the code for a more complex special purpose watch, with the special functions removed. There is plenty of room in the 6351 for lots of extra functions, and to implement just a watch and stop-watch as in this example, the 6351 would not be an appropriate choice. The purpose of the example is to show some real code for a 4 bit microcontroller, and give the reader a feel for some of the methods employed for coping with the peculiarities of such a chip. It also shows some of the routines of earlier chapters in context, in specific code. The 6351 was chosen as a middle range part without the restrictions of some of the smaller parts, but not having the freedom with internal registers and memory access that is common in 8 and 16 bit microcontrollers, and that is also available in many of the top end 4 bit parts.

The 6351 allows interrupt handling, and subroutine calls, and a good selection of conditional branches. As well as allowing some inputs to generate interrupts, some are able, optionally, to release the HALT state. There is only one internal register in the normal sense, that being the accumulator. Even the use of the accumulator is somewhat restricted. Further registers, used for addressing the RAM, and specifying the performance of various functions on the chip, are addressed as I/O ports. Other data stores referred to as registers are locations in the internal RAM, of which there are 1k × 4 bits.

The specification for this product requires that it should display time of day, in 12 or 24 hour format, and should also operate as a stop-watch, giving times to one tenth of a second, and having a lap time facility. The watch also has an alarm facility, giving the user one setable daily alarm time. The functions of the watch are divided into two 'programs', time of day and stop-watch. In time of day there are three modes of operation – normal time display, time setting, and alarm setting. To keep track of which state the watch is in, two nibbles of RAM are used, one called 'PROG' to record which program the watch is in, and the other called 'MODE' to record which mode is in use. The chip drives an LCD consisting of a single row of six digits, and some annunciators. (The digits are driven via a look-up table using the DSPF instruction, whereas the annunciators are driven directly, using the DSP instruction.) The display is accessed via a 16 × 16 bit array. The 16 bits at each address are divided into lower and higher bytes. The DSP and DSPF instructions write to the lower byte at the stated address, and the DSPH and DSPFH instructions address the higher bytes of the array. In this implementation there is provision made to drive 20 segments of annunciators, although only four of them are used here.

The specification, given in brief above, needs to be spelled out in more detail before design can begin. Here is a table showing the functions of the three buttons in the various programs and modes available.

Prog \ Mode	0	1	2
0	Normal Time of Day Button 1 – to Stop-watch Button 2 – to Set Time Button 3 – 12/24 hour	Set Time of Day advance to next field To Set Alarm Increment flashing number	Set Alarm advance to next field To Time of Day Increment flashing number
1	Button 1 – to Time of Day Button 2 – Lap time Button 3 – Start Stop Reset	–	–

In the stop-watch program, button 2 will obviously only record a lap time when the stop-watch is running. The lap time will be displayed for 7 seconds, after which the display will revert to the normal stop-watch display. If the stop-watch is not running, button 2 will do nothing. Button 3 normally toggles between start and stop, but if held down for 1.5 seconds or more, will reset the stop-watch. The stop-watch will continue to run in 'background' while the display is showing Time of Day if button 1 is pressed when the stop-watch is running.

There is no Date facility in this specification, so Time setting is limited to hours, minutes and seconds. For both hours and minutes, in both Time of Day setting and Alarm setting, the selected field flashes, and the number advances by one on pressing button 3, or, if button 3 is held for more than 1.5 seconds, the numbers auto-advance at a rate of 8 per second. In Time of Day, the seconds can be zeroed by pressing button 3 when the seconds are flashing. At the same time, the time will be adjusted to the nearest minute, i.e. if the seconds are between 30 and 59 (inclusive) when the button is pressed, the Time of Day will be advanced by one minute. This simplifies correcting the watch to a time signal, allowing it to be done by a single button press.

When setting the alarm, on entry to the Set Alarm mode, the alarm annunciator flashes. Pressing button 3 will now toggle the annunciator on or off, and the state of the annunciator reflects the state of the alarm, set or not set. The time at which the alarm will sound if it is set is then adjusted in the same way as setting the hours and minutes of Time of Day.

The watch is changed between displaying the time in 12 hour and 24 hour form by pressing button 3 when in the normal Time of Day screen.

The specification given here is very much a functional specification, giving the designer a framework within which to operate, but imposing few restrictions on how the specification is to be implemented on the chosen microcontroller.

The first stage of the design process is to plan out the basic architecture that is to be used for the watch, both hardware and software. There is very little hardware involved – just the display and the three buttons. For the display, this microcontroller is capable of driving 232 display segments in a four way multiplexed display, and only 46 segments are required in this display, so there is a considerable overcapacity. The reason is that in the device from which this example is taken there was a requirement to display much more data, and hence the need for this display capacity. The display drive lines are dedicated to that purpose, and so operation of the display has no effect on other aspects of the circuit. Port 1 was selected as a suitable port to use as inputs from the buttons. It is a general purpose port where all the bits can be set to release the HALT state when the line goes high. There are also built in pull down resistors which can be enabled under program control, but their typical value of 10k ohms makes them rather low for a battery operated device. It would probably be better to use external pull down resistors, of higher value, to avoid effects such as the display dimming when a button is pressed.

As far as the software architecture is concerned there are a number of decisions to be made. The first concerns the timer interrupt. The options available for this microcontroller are at frequencies of 1 Hz, 16 Hz, 32 Hz, and 256 Hz. As far as the Time of Day is concerned, 1 Hz should be frequent enough, as it is not intended to display the Time of Day to any greater precision. This will not be frequent enough for other functions of the watch, but the timer interrupts can be enabled and disabled under program control, and so whatever is needed can be set in motion as and when it is required. The use of a 1 Hz interrupt rate when the watch is not required to do anything except keep track of, and display, the Time of Day, means that the power consumption can be minimised. For the stop-watch, where timing is required to tenths of a second, a 32 Hz interrupt rate was chosen. This does not give the tenths exactly, but the error can be kept to within 12.5 milliseconds. This is an example of the designer interpreting the specification. The specification should have given an accuracy figure for the timing, but in the absence of any such guidance, the designer is free to make what are thought to be appropriate decisions. If the watch was designed to give a greater degree of accuracy than this it is likely that hundredths of seconds would have been required. For the remaining functions of the watch, such as flashing digits, timing button presses, and debouncing the buttons, the interrupt rate of 16 Hz was chosen as the most suitable.

The form of the software then begins to take shape. The timer interrupts are used to keep track of time, initiate display updates, and controlling other functions which are not directly initiated by the user. The buttons are used to release the controller from the HALT state in order to perform whatever function is required. The plan of the software is as follows - after a section which initialises the controller on power up, comes the main loop. Here the controller waits in the HALT state until woken up by an interrupt or button press. When so woken, it proceeds to examine the flags set by the interrupts, and the state of the buttons, in order to establish which operating routines need to be called. The main loop also updates the annunciator display, as the operating routines just set and clear bits in memory to indicate that annunciators should be on or off, and the actual display update takes place the next time the controller program passes through that section of the loop. Following the main loop comes the timer interrupt routine, with sections to be obeyed or skipped according to the type of timer interrupt that has occurred. Provision is made after the timer interrupt routine for a routine to service an external interrupt, although there is no requirement for an external interrupt in this program. This provision is made for demonstration purposes.

After the interrupt routines come the routines which are obeyed when a button is pressed. These make up a substantial portion of the code. The routine BUTTON is called if, in the main loop, it is found that a button has been pressed. Note that it is only obeyed the first time for each button press. Although the main loop may be exercised many times while the button is pressed, the BUTTON routine will only operate once, using the 'button seen' flag to inhibit further passes through the routine. If the alarm is sounding the button press must silence the alarm and do nothing else, but provided that condition does not apply, the BUTTON routine has to make a series of choices, according to which button has been pressed, and what Program and Mode the watch is operating in. The labels for the various sections of the routine indicate the choice that has been made. The numeric portion of the label indicates button, Program and Mode respectively, so for example BT102 will be obeyed when button 1 is pressed when the watch is in Program 0 (Time of Day) and Mode 2 (Alarm setting).

After the BUTTON routine comes a selection of routines which may be called from the BUTTON routine, or from the main program loop. The first of these, LPA (long press action), is obeyed when a button is held down for more than 1.5 seconds. With some keys, the 'long press action' is to repeat the original action, as for example in auto advance of numbers in setting the time. In this case only the normal long press action flag (bit 2 of flag 11) will be set, and the LPA routine can transfer to the BUTTON routine for the action, having first ensured that it is time for the action to be repeated at the 8 Hz rate required. If the 'special long press action' flag is set, (bit 3 of flag 11) then a special section of code is obeyed to perform the special action required. The only special long press action in this specification is for button 3 to clear the stop-watch, but clearly, if the specification called for others they could be put in here.

The routine FLASH controls the flashing functions of the display, flashing the numbers which are to change in time setting, and the stop-watch annunciator if the stop-watch is running. DISEC follows, which is called each time the seconds change, its purpose being to return the stop-watch to displaying the normal elapsed time after a lap time has been displayed for 7 seconds. The following routine,

DISTIM, is responsible for displaying the Time of Day. Note that if the watch is set to 24 hour type display, then the conversion is made at the time of display rather than storing the time in either form, and adding complications to the routine which advances the time. ALCK is called whenever the minutes change, and it compares the current time with the alarm time if the alarm is set, and initiates sounding the alarm if the times are equal. BLKDSP is a routine used to blank the numeric portion of the display. EOPAC (end of press action) is called when a button is released. In some specifications for equipment like this there may be actions to perform on releasing the button, for example, a lap time could have been displayed as long as the button was held down, in which case, releasing the button would signal that the display was to return to normal. In this case, all that has to be done is to check if the 16 Hz interrupt can be switched off, which is handled by the routine NO16HZ. CLPGE is used to clear a page (16 nibbles) of the memory. Note that this routine has several entry points, and can be used to clear just the upper portion of the page. The remaining routines, CADLPT, COPTIM, CST, and DIFF, are used when calculating and displaying the lap time, being called from BT210.

The final section of the code consists of look up tables. These are used for controlling the display, and for the melody generator. For the display there are two tables, one providing numeric displays and the other a selection of alphabetic displays. The alphabetic displays are not in fact needed in this specification. The spaces otherwise not required are filled in with melody tables. In this case the 'melodies' are single tone bleeps, designed to be used with a ceramic sounder which resonates at 4 kHz. If a sounder is driven at a frequency where it does not resonate it makes much less sound, and so extra circuitry is required to boost the drive to the sounder if it is required to produce anything but a single tone. The actual numbers which go into the table to define the melody are coded according to the information provided in the manufacturer's data book.

So here is the code as it might be found in the source file, ready to be assembled.

```
;               Watch / Stop-watch on OKI 6351 chip.
;
                            CHIP      6351
                            LABEL     Watch
                            BATTERY 3.0V
```

```
;Define the I/O ports. The mnemonics used for the ports are more easily
;understood than the port numbers.

PORT0    =    00    ;I/O port, not used
PORT1    =    01    ;I/O port, used for key inputs
PORT2    =    02    ;I/O port, spare
PORT 3   =    03    ;I/O port, output for PROG
PORT 4   =    04    ;I/O port, output for MODE
P00C     =    05    ;Port 0 bit 0 control
P01C     =    06    ;Port 0 bit 1 control
P02C     =    07    ;Port 0 bit 2 control
P03C     =    08    ;Port 0 bit 3 control
P1XC     =    09    ;Port 1 control
P2XC     =    0A    ;Port 2 control
P3XC     =    0B    ;Port 3 control
P4XC     =    0C    ;Port 4 control
SBFF     =    0D    ;Serial data buffer
SCNT     =    0E    ;Serial port control
SCND     =    0F    ;Serial port condition
IRQEX    =    10    ;Interrupt request flags
EIRT     =    11    ;Real Time Interrupt enable
IRQRT    =    12    ;Real Time Interrupt request
IEXM0    =    13    ;External Interrupt 0 control
IEXM1    =    14    ;External Interrupt 1 control
IEXM2    =    15    ;External Interrupt 2 control
TMOUT    =    16    ;Time base 1Hz to 8Hz
CAPTR    =    17    ;Time capturing register
FLAG     =    18    ;Condition Flags
WDOG     =    19    ;Watch-dog timer
FRMT     =    1A    ;Format register - addresses
                    ;lookup table for LCD display
LCDCT    =    1B    ;LCD display control
TEMPO    =    1C    ;Melody Tempo register
BANK     =    1D    ;Memory bank register
PAGE     =    1E    ;Memory page register
WORK     =    1F    ;Working specification register
                    ;used in one RAM addressing mode

;          RAM Allocation

;It is useful to have some mnemonics with which to refer to some locations in
;RAM. In this example the Working specification register was normally set to
```

```
;zero, except during interrupt handling. This made it much simpler to ensure
;correct operation of the interrupts, and gave the interrupt routines access to
;the first 256 nibbles of RAM. The trailing 0 indicates to the assembler that
;working specification mode is in use.

;               Working specification register 0

PROG    =   00      ;'Program' - time of day or stop-watch
MODE    =   10      ;Mode of operation

;Flag registers:

FL1     =   20      ;bit 0 button press seen
                    ;bit 1 button released
                    ;bit 2 time for long press
                    ;bit 3 long press action taken
FL2     =   30      ;bit 0 spare
                    ;bit 1 flash time
                    ;bit 2 flash on
                    ;bit 3 alarm on
FL3     =   40      ;bit 0 16kHz interrupt occurred
                    ;bit 1 alarm sounding
                    ;bit 2 pm
                    ;bit 3 alarm pm
FL4     =   50      ;bit 0 advance minutes
                    ;bit 1 spare
                    ;bit 2 24 hour mode
                    ;bit 3 spare
FL5     =   60      ;bit 0 seconds have changed
                    ;bit 1 spare
                    ;bit 2 displaying stop-watch
                    ;bit 3 stop-watch running
FL6     =   70      ;bit 0 spare
                    ;bit 1 spare
                    ;bit 2 stop-watch reset
                    ;bit 3 spare
FL7     =   80      ;bit 0 stop-watch seconds changed
                    ;bits 1 to 3 spare
FL8     =   90      ;bit 0 spare
                    ;bit 1 end of press action
                    ;bits 2 & 3 spare
FL9     =   A0      ;spare
FL10    =   B0      ;spare
```

```
FL11      =   F0     ;bits 0 & 1 spare
                     ;bit 2 long press action
                     ;bit 3 special long press action
TEMP      =   C0     ;temporary store
TEMP2     =   D0     ;another temporary store

FLNR      =   E0     ;flash number

;         working register 01

;             40     ;32Hz count lsn
;             50     ;32Hz count msn

RXBC      =   80     ;relax back counter
;             90     ;button
PTC0      =   A0     ;press time count (lsn)
PTC1      =   B0     ;press time count (msn)
;             D0     ;save ACC
;             E0     ;timer interrupt IRQ
;             F0     ;save flags

;The remainder of the RAM is not addressed by mnemonics, but by absolute
;addresses. It is mapped as follows: (unallocated memory omitted)

;020  Time seconds units          021  time seconds tens
;022  time minutes units          023  time minutes tens
;024  time hours units            025  time hours tens

;042  Alarm minutes units         043  alarm mins tens
;044  alarm hours units           045  alarm hours tens

;                                 04B  Annunciators 0
;04C  Annunciators 1              04D  Annunciators 2
;04E  Annunciators 3              04F  Annunciators 4

;Annunciator usage:               4D bit 0 = PM
                                  4E bit 2 = Alarm
                                  4E bit 1 = Stop-watch
                                  4F bit 0 = Colon

;060  Stop-watch secs tenths      061  stop-watch secs units
;062  stop-watch secs tens        063  stop-watch mins units
;064  stop-watch mins tens        065  stop-watch hours units

;Now here is the start of the code proper, starting from address 0, the reset
;vector.
```

```
        ORG     0

;The first section of code initialises the microcontroller in the desired mode of
;operation.

INIT:   OUT     #0,WORK         ;set working specification
                                ;register to 0
        OUT     #4,LCDCT        ;quadriplexed LCD
        OUT     #2,P00C         ;no clock o/p, P00 not release
                                ;HALT, Hi Z, i/p
        OUT     #A,P01C         ;Port 0.1, input Hi Z, BUFF = 1

;(BUFF controls a voltage halver circuit, when set to 0 reduces power
;consumption, but also reduces voltage tolerance.)

        OUT     #1,P02C         ;P02 as o/p
        OUT     #6,P1XC         ;Port 1 as input, HALT release, Hi Z

;the 3 buttons used to control the watch are input to port 1.

        OUT     #A,P2XC         ;port 2 as input, no HALT rel.
                                ;Hi Z, serial o/p also Hi Z

;Because the spare I/O ports 0 and 2 are configured as input ports they
;should be externally connected to 0 volts.

        OUT     #1,P3XC         ;port 3 as o/p
        OUT     #1,P4XC         ;port 4 as o/p
        OUT     #C,SCNT         ;serial clock as input
        OUT     #1,EIRT         ;enable 1 sec interrupts
        OUT     #1,TEMPO        ;fastest melody tempo
        OUT     #4,CAPTR        ;capture 1kHz to 128Hz
        OUT     #C,FRMT         ;table C for LCD (numerics)

;The memory is cleared, from 0 to 3FF, using the bank and page registers.

        MOV     #0,TEMP         ;clear memory
        MOV     #0,TEMP2
INI1A:  OUT     TEMP,BANK
INI1B:  OUT     TEMP2,PAGE
        CALL    CLPGE           ;clear all 16 nibbles of the current
                                ;page
        INC     TEMP2
        BCC     INI1B           ;carry set when 16 pages of current
                                ;bank cleared (BCC = branch if carry
                                ;clear)
```

```
        INC     TEMP
        CMP     #4,TEMP
        BLT     INI1A         ;branch if TEMP less than 4

        MOV     #4,FL6        ;stop-watch reset
        OUT     #2,PAGE       ;init time of day
        OUT     #0,BANK
        MOV     #2,4P
        MOV     #1,5P         ;12:00 am
        OUT     #4,PAGE
        MOV     #2,4P         ;set alarm to 12.00 am
        MOV     #1,5P
        BIC     #C,FL3        ;clear pm bits
        OUT     #0,TMOUT
        CALL    LPAF          ;reset stop-watch
        OUT     #4,PAGE
        BIC     #1,EP         ;stop-watch anun OFF
```

;In the main program loop, various flags are examined and service routines ;called if required. The inputs from the buttons are also polled, and if any is ;pressed the button servicing routine is called. The flags examined are set in ;interrupt routines, or in some cases, by the service routines.

```
LOOP:   HALT
        NOP
        OUT     #A,WDOG       ;kick the watch-dog
        OUT     #5,WDOG
        OUT     #4,PAGE
        MOV     BP,ACC        ;display annunciators
        DSPH    0,CP          ;DSPH drives directly the higher
        MOV     DP,ACC        ;8 bits at a given address with
        DSPH    1,EP          ;the data from the RAM
        MOV     FP,ACC        ;register and the Accumulator.
        DSPH    2,FP
        OUT     PROG,PORT3    ;Ports 3 & 4 were not needed, so
        OUT     MODE,PORT4    ;the state of the watch was output,
                              ;for possible use by external circuitry.
L1:     BIT     #2,FL2        ;flash time?
        BZE     L2            ;branch if not.
        CALL    FLASH
L2:     BIT     #4,FL1        ;time for long press action?
        BZE     L3
```

```
        CALL   LPA
L3:     INP    PORT1,TEMP   ;any buttons pressed?
        CMP    #0,TEMP
        BZE    L4
        CALL   BUTTON
L4:     BIT    #1,FL4       ;advance minutes?
        BZE    L8
        CALL   BT301A       ;advance time by one minute
        CALL   ALCK         ;check alarm time
        OUT    #1,PAGE
        CMP    #0,PROG      ;watch should not stay in setting
        BNE    L8           ;modes if buttons not pressed for 3
                            ;minutes.
        DEC    8P           ;relax back counter
        BCC    L8
        CALL   BT202        ;return to MODE = 0
        OUT    #1,PAGE
L8:     BIT    #1,FL5       ;seconds changed?
        BZE    L11
        CALL   DISEC
L11:    BIT    #2,FL8       ;end of press action?
        BZE    L12
        CALL   EOPAC
L12:    CMP    #0,PROG
        BNZ    L13
        CALL   DISTIM       ;display time of day if PROG=0
L13:    JMP    LOOP

;Timer Interrupt Routine:

TIMI:   OUT    #1,WORK      ;timer interrupt
        MOV    ACC,D0       ;save ACC
        INP    FLAG,F0      ;save flags
        INP    IRQRT,E0     ;which interrupt?
TIM32:  BIT    #4,E0        ;32Hz?
        BNZ    TIM320
        JMP    TIM16
TIM320: INC    40           ;stop-watch interrupt
                            ;increment 32Hz tick count
;32Hz does not give tenths exactly. Tenths must be incremented on ticks 0, 3,
;6, 10, and 13 for minimum error.
        BZE    TIM321
```

```
        CMP     #3,40       ;find times to inc 1/10ths
        BEQ     TIM321
        CMP     #6,40
        BEQ     TIM321
        CMP     #A,40
        BEQ     TIM321
        CMP     #D,40
        BEQ     TIM321
        JMP     TIM32Z
TIM321: OUT     #6,WORK
        SEC                 ;set carry
        ADCA    00          ;add one to stop-watch time
        ADCA    10          ;ADCA adds carry, base 10
        ADC6    20          ;ADC6 adds carry, base 6
        ADCA    30
        ADC6    40
        ADCA    50
        OUT     #0,WORK
        BIT     #4,FL5      ;displaying stop-watch?
        BZE     TIM32Z
        OUT     #6,WORK
        DSPF    1,00        ;if so, output stop-watch
        DSPF    2,10        ;time to display
        DSPF    3,20
        DSPF    4,30
        CMP     #0,40
        BEQ     TIM322
        DSPF    5,40
        JMP     TIM32Z
TIM322: OUT     #1,WORK
        MOV     #F,TEMP
        DSPF    5,TEMP      ;blank leading zero
TIM32Z: OUT     #1,WORK

;16Hz interrupts are used for servicing the buttons and for flashing digits
;when required.

TIM16:  BIT     #2,E0       ;16Hz?
        BNZ     TIM160
        JMP     TIM1
TIM160: INP     PORT1,TEMP
        CMP     #0,TEMP     ;button still pressed?
```

```
          BEQ     TIM16N        ;branch if not
;increment press time count if button still pressed.
          INC     PTC0          ;press time count lsn
          BCC     TIM16A
          INC     PTC1          ;press time count msn
TIM16A:   OUT     #0,WORK
          BIS     #1,FL3        ;set 16kHz interrupt flag
          BIC     #2,FL1        ;clear button released flag
          OUT     #1,WORK
          CMP     #8,PTCO       ;1.5 seconds?
          BNE     TIM16Z
          CMP     #1,PTC1
          BNE     TIM16Z
          OUT     #0,WORK
          BIT     #8,FL1        ;long press action taken?
          BNZ     TIM16Z        ;branch if so
          BIS     #4,FL1        ;time for long press action
          JMP     TIM16Z
TIM16N:   OUT     #0,WORK
          BIT     #2,FL1        ;button released?
          BNZ     TIM160        ;branch if so
          BIS     #2,FL1        ;set flag
          JMP     TIM16Z
TIM16O:   BIT     #1,FL1        ;button press seen?
          BZE     TIM16Q        ;branch if not
          BIC     #D,FL1        ;End of press - clear flags
          BIS     #2,FL8        ;end of press action flag
          OUT     #4,P1XC       ;enable halt release (buttons)
TIM16Q:   OUT     #1,WORK       ;No buttons active
          INC     PTC0
          BIT     #1,PTC0       ;flash time?
          BNZ     TIM16Z
          BIT     #2,PTC0
          BNZ     TIM16Z
          OUT     #0,WORK
          BIS     #2,FL2        ;set flash time flag
TIM16Z:   OUT     #1,WORK

;1Hz interrupt used to keep track of Time of Day.

TIM1:     BIT     #1,E0         ;1Hz?
          BNZ     TIM10
```

```
        JMP     TIMZ
TIM10:  OUT     #2,WORK
        SEC
        ADCA    00          ;decimal increment seconds
        ADC6    10          ;increment seconds tens if carry
        OUT     #0,WORK
        BCC     TIM12       ;branch if no carry
        BIS     #1,FL4      ;set 'advance mins' bit
TIM12:  BIS     #1,FL5      ;seconds have changed
        BIT     #2,FL3      ;alarm sounding?
        BZE     TIMZ        ;branch if not
        MSA     B3          ;sound alarm
TIMZ:   OUT     #1,WORK
        MOV     D0,ACC      ;restore ACC
        OUT     F0,FLAG     ;restore flags
        OUT     #0,WORK
        RTI

EXTI:   OUT     #1,WORK     ;External interrupt routine
        MOV     ACC,D0      ;save ACC
        INP     FLAG,F0     ;save Z, C, G
        INP     IRQEX,TEMP  ;check it was correct
        BIT     #1,TEMP     ;                  interrupt
        BZE     TIMZ

;No external interrupt is used in this program, but it could be included here if
;required.

        JMP     TIMZ

;The button service routine:

BUTTON: BIT     #1,FL1      ;flags 1 bit 1 = button seen
        BNZ     BUTZ        ;return if seen
        BIS     #1,FL1      ;set button seen flag
        OUT     #0,P1XC     ;disable halt release
        OUT     #1,PAGE
        MOV     #3,8P       ;set relax back to 3 mins
        MOV     #0,AP
        MOV     #0,BP       ;zero press time count
        MOV     TEMP,ACC    ;store button code
        MOV     ACC,9P      ;(was put in TEMP at L3)
        INP     EIRT,TEMP2
```

```
        BIS     #2,TEMP2        ;enable 16Hz interrupts
        OUT     TEMP2,EIRT
        BIT     #2,FL3          ;alarm sounding?
        BZE     BUT0            ;branch if not
;any button will silence the alarm, in which case it does not perform its usual
;function.
        BIC     #2,FL3          ;cancel alarm
        RET
BUT0:   OUT     #2,PAGE
        BIC     #C,FL11         ;clear long press action flags
        BIT     #8,TEMP         ;TEMP contains button code.
        BNZ     BUT1            ;button 1
        BIT     #4,TEMP
        BNZ     BUT2            ;button 2
        BIT     #2,TEMP
        BNZ     BUT3            ;button 3
BUTZ:   RET
BUT1:   OUT     #4,PAGE
        JMP     @PROG
        JMP     BT10            ;button 1, time of day
        JMP     BT11            ;button 1, stop-watch

BUT2:   BIS     #C,FL11         ;special long press action
        JMP     @PROG
        JMP     BT20
        JMP     BT21

BUT3:   JMP     @PROG
        JMP     BT30
        JMP     BT31

BT10:   JMP     @MODE           ;Button 1 in Prog = 0
        JMP     BT100           ;normal time of day screen
        JMP     BT101           ;set ToD

BT102:  INC     FLNR            ;set alarm

;When setting alarm time, button 1 determines which numbers will change.
;FLNR tells the FLASH routine what to flash, and the button 3 routine what
;to change. FLNR = 0 for Alarm on / off, FLNR = 1: flash / change hours,
;FLNR = 2: flash / change minutes, FLNR = 5 for Alarm on / off without flashing
;symbol.
```

```
        CMP     #3,FLNR
        BLT     BT102Z
        MOV     #1,FLNR
        BGT     BT102Z
        MOV     #0,FLNR
BT102Z: JMP     BT3022          ;correct alarm annunciator

BT100:  MOV     #1,PROG         ;change to stop-watch

;When showing normal Time of Day, button 1 moves the watch to Stop-watch.

BT100A: BIS     #4,FL5          ;displaying stop-watch
        OUT     #4,PAGE
        BIS     #1,EP           ;stop-watch symbol on
        BIS     #2,FP           ;colon ON
        BIC     #1,DP           ;pm off
        OUT     #6,PAGE
BT100B: CALL    BLKDSP          ;blank display (not annuns)
    ;display stop-watch time:
        DSPF    1,0P            ;tenths
        DSPF    2,1P            ;seconds units
        DSPF    3,2P            ;seconds tens
        DSPF    4,3P            ;minutes units
        MOV     4P,ACC          ;suppress leading zero
        JMP     DSTM02

BT101:  INC     FLNR

;When setting time of day, button 1 selects which number to change, as in
;alarm setting, using FLNR. Here, when FLNR = 0 the seconds are zeroed.
;With FLNR = 1 or 2 the hours or minutes flash / change respectively.

        CMP     #3,FLNR
        BLT     BT101Z
        MOV     #0,FLNR
BT101Z: RET

;Other programs could be included to add extra functions.

BT11:   MOV     #0,PROG         ;Return to Time of Day
        MOV     #0,MODE
        BIC     #1,EP           ;stop-watch symbol off
        RET

;In Time of Day, button 2 selects mode of operation:
```

```
;MODE = 0 : Time of Day, MODE = 1 : set time,
;MODE = 2 : Set Alarm.

BT20:    JMP     @MODE       ;button 2 in t o d
         JMP     BT200
         JMP     BT201
BT202:   MOV     #0,MODE     ;Go back to Time of Day
         MOVD    ACC,610     ;cancel flashing
         JMP     BT3022      ;correct alarm annunciator

BT200:   MOV     #1,MODE     ;Set time
BT200A:  MOV     #0,FLNR
BT200B:  OUT     #1,PAGE
         MOV     #F,6P       ;flashing active
         RET
BT201:   MOV     #2,MODE     ;Set alarm
         JMP     BT200A

BT21:    BIT     #8,FL5      ;stop-watch running?
         BNZ     BT210       ;branch if so
         RET
BT210:   CALL    COPTIM      ;copy current time
         CALL    CADLPT      ;calculate & display lap time
         MOVD    660,ACC     ;copy current time
         MOV     ACC,6P
         MOVD    760,ACC
         MOV     ACC,7P
         MOVD    860,ACC
         MOV     ACC,8P
         MOVD    960,ACC
         MOV     ACC,9P
         MOVD    A60,ACC
         MOV     ACC,AP
         RET

;The interrupt vectors are fixed at 400 and 401.

         ORG     400
         JMP     TIMI        ;Timer interrupts

         ORG     401         ;External Interrupts
         JMP     EXTI

BT30:    OUT     #2,PAGE
```

```
          JMP     @MODE        ;button 3, time of day prog.
          JMP     BT300        ;toggle 12/24 hour
          JMP     BT301        ;mode 1 - set time of day
BT302:    OUT     #4,PAGE      ;mode 2 - set alarm time
          BIS     #4,FL11      ;allow long press action
          JMP     @FLNR
          JMP     BT3020
          JMP     BT3011
          JMP     BT3012
          RET
          RET
BT3020:   XOR     #8,FL2       ;toggle daily alarm on/off
          MOV     #5,FLNR      ;cancel flashing
BT3022:   OUT     #4,PAGE
          BIS     #4,EP        ;annunciator on
          BIT     #8,FL2
          BNZ     BT3021
          BIC     #4,EP        ;annunciator off
BT3021:   RET

BT301:    BIS     #4,FL11      ;allow long press action
          JMP     @FLNR
          JMP     BT3010       ;change seconds
          JMP     BT3011       ;change hours
BT3012:   MOV     #0,TEMP      ;change minutes
          SEC
          ADC     2P
          BCC     BT301X
          ADC6    3P
BT301X:   DSPF    2,2P
          DSPF    3,3P
          RET

BT3010:   MOV     1P,ACC
          OUT     #0,TMOUT     ;zero seconds
          MOV     ACC,TEMP
          MOV     #0,0P
          MOV     #0,1P
          DSPF    0,0P
          DSPF    1,1P
          CMP     #2,TEMP      ;seconds > 29?
          BGT     BT301A
```

```
BT301B:  RET
BT301A:  BIC     #1,FL4        ;clear change minute flag
         OUT     #2,PAGE
         MOV     #0,TEMP
         SEC
         ADC     2P            ;advance minutes
         BCC     BT301B
         ADC6    3P
         BCC     BT301B
         ADC     4P
         BCC     BT301C        ;advance hours
         ADC     5P
BT301C:  CMP     #1,5P         ;check for 12 & 13
         BLT     BT301B
         CMP     #2,4P
         BLT     BT301B
         BGT     BT301D
         XOR     #4,FL3        ;toggle pm if hours = 12
         RET
BT301D:  MOV     #1,4P         ;set hours = 01 if > 12
         MOV     #0,5P
         RET
BT3011:  MOV     #0,TEMP
         SEC
         ADC     4P
         BCC     BT301Z
         ADC     5P
BT301Z:  CMP     #1,5P         ;check for 13
         BNE     BT301Y
         CMP     #3,4P
         BNE     BT301V
         MOV     #1,4P
         MOV     #0,5P
BT301Y:  JMP     DSTM00
BT301V:  CMP     #2,4P         ;toggle pm at 12.00
         BNE     BT301Y
         CMP     #2,MODE       ;alarm set?
         BNE     BT301W
         XOR     #8,FL3        ;toggle pm
         JMP     DSTM00        ;display hours with leading
BT301W:  XOR     #4,FL3        ;               zero suppression
```

```
        JMP     DSTM00

BT300:  XOR     #4,FL4        ;toggle 12/24 hour
        BIC     #C,FL11       ;cancel long press action
        RET

BT31:   OUT     #8,PAGE
        BIS     #C,FL11       ;special long press action
        BIT     #8,FL5        ;stop-watch running?
        BNZ     BT310         ;branch if so
        JMP     BT311         ;jump if not
BT310:  BIC     #8,FL5        ;stop stop-watch
BT310F: INP     EIRT,TEMP     ;disable 32Hz interrupts
        BIC     #4,TEMP
        OUT     TEMP,EIRT
        CALL    NO16HZ
BT310E: JMP     BT100A        ;display stop-watch time

BT311:  BIS     #8,FL5        ;start stop-watch
        INP     EIRT,TEMP
        BIS     #6,TEMP
        OUT     TEMP,EIRT     ;enable 16Hz & 32Hz interrupts
        BIC     #8,FL6        ;not last lap
        OUT     #1,PAGE
        MOV     #0,4P         ;clear 32Hz int count
        MOV     #0,5P
        BIT     #4,FL6        ;starting from reset?
        BZE     BT311A
        BIC     #4,FL6        ;clear reset flag
BT311A: JMP     BT100A        ;display stop-watch time

;Long press action. When a button is held, 'long press action' must take place
;after 1.5 seconds if the flag (FL11 bit 2) is set. Normally the action is to repeat
;the original action, as in setting time to advance quickly through the numbers.
;If FL11 bit 3 is also set, the action is a special one, as in resetting the stop-
;watch.

LPA:    OUT     #1,PAGE       ;long press action
        BIT     #8,FL1        ;action already taken?
        BZE     LPAA          ;branch if not
LPARET: RET
LPAA:   BIT     #4,FL11       ;action expected?
        BZE     TOLPAZ        ;exit if not
```

```
         INP     PORT1, TEMP  ;get keypress
         BIT     #8,FL11      ;special action?
         BNZ     LPAB         ;branch if so
         BIT     #1,AP        ;PTC0, press time count
         BNZ     LPARET       ;repeat at 8Hz
         BIT     #1,FL3       ;16kHz interrupt flag
         BZE     LPARET
         BIC     #1,FL3
         JMP     BUT0         ;repeat button action

;Special actions:

LPAB:    BIT     #4,TEMP      ;button 3?
         BZE     LPAC         ;branch if button 3
TOLPAZ:  JMP     LPAZ

LPAC:    JMP     @PROG        ;Button 3 special long press action
         RET                  ;no action in Time of Day

LPAF:    OUT     #6,PAGE      ;reset stop-watch
         CALL    CLPGE
         DSPF    1,0P
         DSPF    2,1P
         DSPF    3,2P
         DSPF    4,3P
         MOV     #F,TEMP
         DSPF    5,TEMP
         BIS     #4,FL6       ;stop-watch reset flag ON
         BIC     #8,FL5       ;stop-watch not running
         INP     EIRT,TEMP
         BIC     #4,TEMP      ;disable 32Hz interrupts
         OUT     TEMP,EIRT
         OUT     #7,PAGE
         CALL    CLPGE        ;clear split time & lap no.
         OUT     #4,PAGE
         BIS     #1,EP        ;stop-watch anun ON
LPAZ:    BIC     #4,FL1       ;clear time for action flag
         BIS     #8,FL1       ;set action taken flag
         RET

FLASH:   BIC     #2,FL2       ;Flash time – clear flag
         XOR     #4,FL2       ;toggle on/off flag
         OUT     #4,PAGE      ;address annunciators
```

```
        MOV    ACC,TEMP     ;FL2 to TEMP
        BIC    #B,TEMP      ;clear all but FLASH bit
        ASR    TEMP         ;FLASH bit to bit 0
        ASR    TEMP
        BIT    #8,FL5       ;stop-watch running?
        BZE    FLRZ1A
        XOR    ACC,EP       ;toggle stop-watch annun
FLRZ1A: CMP    #1,PROG
        BLT    FLASHA
FLBRET: RET

FLASHA: BIT    #4,FL2       ;on or off?
        BNE    FLASH1       ;branch if off
        OUT    #2,PAGE      ;address time of day
        CMP    #2,MODE
        BNE    FLASH0
        OUT    #4,PAGE      ;if MODE = 2, address alarm time
                            ;          instead
FLASH0: JMP    @FLNR
        JMP    FLON0        ;display secs/alarm annun
        JMP    DSTM00       ;display hours
        JMP    FLON2        ;display minutes
        RET
        RET
        RET                 ;FLNR = 5, alarm on/off set
FLON0:  CMP    #2,MODE      ;alarm setting?
        BNZ    FLON01       ;branch if time setting
FLON00: XOR    #4,EP        ;toggle alarm annunciator
        RET
FLON01: DSPF   0,0P         ;display seconds
        DSPF   1,1P
        RET
FLON2:  DSPF   2,2P         ;display minutes
        DSPF   3,3P
FL2RET: RET

;clear flashing element of display if FL2 bit 2 = 0

FLASH1: CMP    #0,MODE
        BEQ    FL2RET       ;no flashing in Mode 0
        MOV    #F,TEMP
        JMP    @FLNR
```

```
         JMP    FLOF0      ;0 - secs / alarm annun.
TOFLF1:  JMP    FLOF1      ;1 - hours
TOFLF2:  JMP    FLOF2      ;2 - minutes
         JMP    FLOF1      ;3 - not used
         JMP    FLOF2      ;4 - not used
         RET               ;5 - no flash
FLOF0:   DSPF   0,TEMP     ;clear seconds
         DSPF   1,TEMP
         CMP    #2,MODE    ;alarm setting?
         BZE    FLON00     ;toggle alarm annun.
         RET
FLOF1:   DSPF   4,TEMP     ;clear hours
         DSPF   5,TEMP
         RET
FLOF2:   DSPF   2,TEMP     ;clear minutes
         DSPF   3,TEMP
         RET

;DISEC routine is called once per second

DISEC:   BIC    #1,FL5     ;clear 'seconds have changed' flag
DS3:     JMP    @PROG
DS31:    RET               ;time of day

DS4:     OUT    #7,PAGE    ;stop-watch
         CMP    #F,5P
         BEQ    DS4RET
         DEC    5P         ;lap display count
         BCS    DS42
DS4RET:  RET               ;leave lap time on display for 7 seconds
DS42:    JMP    BT100A     ;display stop-watch

;DISTIM displays the time of day.

DISTIM:  OUT    #4,PAGE    ;Display time of day
         BIS    #2,FP      ;put on colon
         BIC    #1,DP      ;cancel PM annunciator
         BIT    #4,FL4     ;24 hour?
         BNZ    DSTIM0
         BIT    #4,FL3     ;pm?
         BZE    DSTIM0
         BIS    #1,DP      ;PM annunciator ON
DSTIM0:  JMP    @MODE
```

```
         JMP    DSTM0        ;normal time of day
         JMP    DSTM1        ;time setting
DSTM2:   MOV    #F,TEMP      ;display alarm time
         DSPF   0,TEMP       ;blank seconds
         DSPF   1,TEMP
         BIC    #1,DP        ;PM annunciator OFF
         BIT    #4,FL4       ;24 hour?
         BNZ    DSTM20
         BIT    #8,FL3       ;Alarm pm?
         BZE    DSTM20
         BIS    #1,DP        ;PM annunciator ON
DSTM20:  CMP    #2,FLNR
         BZE    DSTM21
         DSPF   2,2P         ;display minutes
         DSPF   3,3P
DSTM21:  CMP    #1,FLNR
         BZE    DSTM22
         JMP    DSTM00       ;hours display
DSTM22:  RET

DSTM0:   OUT    #2,PAGE
         DSPF   0,0P         ;display T o D
         DSPF   1,1P
         DSPF   2,2P
         DSPF   3,3P
DSTM00:  BIT    #4,FL4       ;hours - 24?
         BNZ    DSTMO4
DSTM06:  DSPF   4,4P
         MOV    5P,ACC
DSTM02:  MOV    ACC,TEMP
DSTM03:  CMP    #0,TEMP
         BNE    DSTMO1
DSTM08:  MOV    #F,TEMP
DSTM01:  JMP    BLKT1        ;suppress leading zero

DSTM04:  CMP    #2,MODE      ;alarm display?
         BNE    DSTM09
         BIT    #8,FL3       ;alarm pm?
         BNZ    DSTM05
         JMP    DSTMOA
DSTM09:  BIT    #4,FL3       ;24 hour time display - pm?
         BNZ    DSTM05
```

```
DSTM0A: CMP    #1,5P       ;AM - 12?
        BNE    DSTM06      ;normal display if not
        CMP    #2,4P
        BNE    DSTM06
        MOV    #0,TEMP
        DSPF   4,TEMP
        JMP    DSTM08
DSTM05: INP    PAGE,TEMP   ;save PAGE
        OUT    #4,PAGE     ;PM - cancel annunciator
        BIC    #1,DP
        OUT    TEMP,PAGE   ;restore PAGE
        CMP    #1,5P       ;12?
        BNE    DSTM07
        CMP    #2,4P
        BEQ    DSTM06      ;normal display for 12
DSTM07: CLC
        MOV    #2,TEMP
        MOV    4P,ACC
        ADC    TEMP        ;hours units plus 2
        DSPF   4,TEMP
        MOV    #1,TEMP
        MOV    5P,ACC
        ADC    TEMP        ;hours tens + 1 + carry
        DSPF   5,TEMP
DSTM13: RET

DSTM1:  OUT    #2,PAGE
        CMP    #0,FLNR     ;display set time
        BZE    DSTM11
        DSPF   0,0P
        DSPF   1,1P        ;Secs - don't display if flashing
DSTM11: CMP    #2,FLNR     ;flashing minutes?
        BZE    DSTM12
        DSPF   2,2P        ;display minutes
        DSPF   3,3P
DSTM12: CMP    #1,FLNR     ;flashing hours?
        BZE    DSTM13
        JMP    DSTM00      ;display hours.

;ALCK Alarm check routine, checks if it is time to sound the alarm

ALCK:   OUT    #4,PAGE
```

```
         BIC     #2,FL3
         BIT     #8,FL2       ;alarm set?
         BZE     ALCKZ        ;RET if not
         MOVD    220,ACC      ;compare Time of Day
         CMP     ACC,2P       ;          with Alarm Time
         BNE     ALCKZ
         MOVD    320,ACC
         CMP     ACC,3P
         BNE     ALCKZ
         MOVD    420,ACC
         CMP     ACC,4P
         BNE     ALCKZ
         MOVD    520,ACC
         CMP     ACC,5P
         BNE     ALCKZ
         MOV     FL3,ACC      ;are PM bits = ?
         MOV     ACC,TEMP
         BIC     #3,TEMP
         BZE     ALCK2
         CMP     #C,TEMP
         BNZ     ALCKZ
ALCK2:   MSA     B3           ;sound alarm
         BIS     #2,FL3
ALCKZ:   JMP     NO16HZ       ;check if 16Hz can be switched off

BLKDSP:  MOV     #F,TEMP      ;blank display
         DSPF    0,TEMP       ;(not annunciators)
         DSPF    1,TEMP
         DSPF    2,TEMP
         DSPF    3,TEMP
         DSPF    4,TEMP
BLKT1:   DSPF    5,TEMP
         RET

EOPAC:   BIC     #2,FL8       ;End of press action - clear flg
         JMP     NO16HZ       ;cancel 16Hz if possible.

CLPGE:   MOV     #0,0P        ;clear page
         MOV     #0,1P
CLRPG2:  MOV     #0,2P
         MOV     #0,3P
         MOV     #0,4P
```

```
        MOV     #0,5P
        MOV     #0,6P
        MOV     #0,7P
CLRPG8: MOV     #0,8P
        MOV     #0,9P
        MOV     #0,AP
CLRPGB: MOV     #0,BP
CLRPGC: MOV     #0,CP
        MOV     #0,DP
        MOV     #0,EP
        MOV     #0,FP
CLRPGR: RET

;Calculate & Display Lap Time - set to Page 7 on entry

CADLPT: CALL    DIFF
CDLA:   CMP     #1,PROG
        BNE     CLRPGR          ;nearest handy RET
        CMP     #0,MODE
        BNE     CLRPGR
        BIC     #4,FL5          ;cancel display stop-watch
        MOV     #6,5P           ;for 7 seconds
        JMP     BT100B          ;display Lap time

COPTIM: OUT     #7,PAGE
        MOVD    060,ACC         ;copy current time
        MOVD    ACC,660
        MOV     ACC,0P
        MOVD    160,ACC
        MOVD    ACC,760
        MOV     ACC,1P
        MOVD    260,ACC
        MOVD    ACC,860
        MOV     ACC,2P
        MOVD    360,ACC
        MOVD    ACC,960
        MOV     ACC,3P
        MOVD    460,ACC
        MOVD    ACC,A60
        MOV     ACC,4P
        RET
```

```
CST:     MOV    6P,ACC      ;copy split time to WORK1
         MOVD   ACC,650
         MOV    7P,ACC
         MOVD   ACC,750
         MOV    8P,ACC
         MOVD   ACC,850
         MOV    9P,ACC
         MOVD   ACC,950
         MOV    AP,ACC
         MOVD   ACC,A50
         RET

DIFF:    MOV    6P,ACC      ;tenths
         CLC
         SBC    0P
         MOV    7P,ACC      ;seconds units
         SBC    1P
         SBC6   2P
         SBCA   3P
         SBC6   4P
         MOV    8P,ACC      ;seconds tens
         SUB    ACC,2P
         BCC    DIFF1
         ADD    #6,2P       ;base 6 adjust
         SBCA   3P
         SBC6   4P
DIFF1:   MOV    9P,ACC      ;minutes units
         CLC
         SBC    3P
         SBC6   4P
         MOV    AP,ACC      ;minutes tens
         SUB    ACC,4P
         BCC    DIFF2
         ADD    #6,4P
DIFF2:   RET

NO16HZ:  OUT    #1,PAGE     ;cancel 16Hz interrupts
         BIT    #8,6P       ;if flashing not active
         BNZ    NO16ZZ
         BIT    #2,FL1      ;and button released
         BZE    NO16ZZ
         BIT    #1,FL1      ;and button press not seen
```

```
        BNZ     NO16ZZ
        BIT     #8,FL5          ;and stop-watch not running
        BNZ     NO16ZZ
        INP     EIRT,TEMP
        BIC     #2,TEMP
        OUT     TEMP,EIRT
NO16ZZ: RET

        ORG     OFDC

;Bleep at 4kHz to resonate with a 4kHz ceramic sounder.
;The code at B3: is a table of data for the MSA instruction.

B3:     DW      0626            ;Double bleep
        DW      0022            ;for Time of Day
        DW      0626            ;alarm
        DW      0082

        ORG     OFE0            ;Character table C
LCDTBO: DW      00FC            ;char 0
        DW      0060            ;1
        DW      00DA            ;2
        DW      00F2            ;3
        DW      0066            ;4
        DW      00B6            ;5
        DW      00BE            ;6
        DW      00E4            ;7
        DW      00FE            ;8
        DW      00F6            ;9
        DW      0000            ;blank A

B2:     DW      067E            ;Long bleep
        DW      067E
        DW      06FE
        DW      0000

        DW      0000            ;blank F

        DW      006E            ;H      0       Table E
        DW      008E            ;F      1
        DW      00EC            ;N      2
        DW      001E            ;t      3
        DW      001C            ;L      4
        DW      00EE            ;A      5
```

```
        DW      0002        ;-      6
        DW      00FC        ;0      7
        DW      009E        ;E      8
        DW      002A        ;n      9
        DW      007A        ;d      A

;Bleeps 1, 2 & 4 are not used by this specification.

B4:     DW      0622        ;double bleep
        DW      002E
B1:     DW      0622        ;single bleep
        DW      0082

        DW      0000

        END
```

Appendix I
Manufacturers' head office addresses

Fujitsu

Fujitsu Ltd
6-1, Marunouchi, 1-chome,
Chiyodo-ku,
Tokyo 100,
Japan
Tel (03) 216-3211
Telex J22833

Hitachi

Hitachi Ltd
Semiconductor & IC Division
Karukozaka MN Bldg.,
2-1, Ageha-cho,
Shinjuku-ku
Tokyo 162
Japan
Tel (03) 266-9359
Fax (03) 235-2375

Mitsubishi

Mitsubishi Electric Corporation,
Semiconductor Marketing
Division,
2-3, Marunouchi 2-chome,
Chiyoda-ku,
Tokyo 100,
Japan
Tel (03) 218-3473
Fax (03) 214-5570
Telex 24532 MELCO J

National Semiconductor

National Semiconductor
Corporation,
2900 Semiconductor Drive,
P.O. Box 58090,
Santa Clara,
CA 95052-8090
USA
Tel (408) 721-5000
TWX (910) 339-9240

NEC

NEC Corporation
NEC Headquarters
7-1 Shiba 5-chome
Minato-ku
Tokyo 108
Japan
Tel (03) 798 6046
Fax (03) 798 6056
Telex 222686

OKI

OKI Electric Industry Co. Ltd.,
7-5-25 Nishishinjuku,
Shinjuku-ku
Tokyo 160
Japan
Tel (03) 5386-8100
Fax (03) 5386-8110
Telex J27662

Panasonic

Matsushita Electronics
Corporation,
Semiconductor Group,
Nagaokakyo,
Kyoto 617,
Japan
Tel (075) 951-8151

Sanyo

Sanyo Electric Co. Ltd.,
Natsume Bldg.,
18-6, 2-chome, Yushima,
Bunkyo-ku,
Tokyo 113,
Japan
Tel (03) 818-1144
Fax (03) 818-1274
Telex J26463 TKYSANYO

Seiko Epson

Seiko Epson Corporation,
IC Sales & Marketing Department,
10F Shinjuku NS Building,
2-4-1, Nishi-Shinjuku,
Shinjuku-ku
Tokyo 163,
Japan
Tel (03) 3348-7367
Fax (03) 3348-7188

SGS-Thomson

SGS-Thomson Microelectronics
7 avenue Gallieni,
BP93,
94253 Gentilly Cedex
France
Tel (01) 47.40.75.75
Fax (01) 47.40.79.10
Telex 632570 STMHQ

Sharp

Sharp Corporation,
International Sales & Marketing
Group,
IC / Electronic Components,
2613-1, Ichinimoto-cho,
Tenri-city,
Nara 632
Japan
Tel (07436) 5-1321
Fax (07436) 5-1532
Telex LABOMETA-B J63428

Texas Instruments

Texas Instruments Inc
P O Box 65474
Dallas
Texas 75265
U S A
Tel (214) 995 2011
Fax (214) 995 4360
Telex 682 9291

Toshiba

Toshiba Corporation
International Operations –
Electronic Components
1-1, Shibaura 1-chome,
Minato-ku,
Tokyo, 105-01,
Japan
Tel (03) 457-3495
Fax (03) 451-0576
Telex J22587

Appendix II
Manufacturers' contact points in the United States

Fujitsu

Fujitsu Microelectronics Inc.
IC Division
3545 N First Street
San Jose
CA 95134
Tel (800) 642-7616
Fax (408) 432-9044

Hitachi

Hitachi America Ltd
Semiconductor & IC Division
2000 Sierra Point Pkwy
Brisbane
CA 94005
Tel (415) 589-8300
Fax (415) 583-4207

Mitsubishi

Mitsubishi Electronics America Inc
1050 E Arques Ave
Sunnyvale
CA 94086
Tel (408) 730-5900
Fax (408) 730-4972

National Semiconductor

National Semiconductor Corp
2900 Semiconductor Drive
Box 58090
Santa Clara
CA 95052-8090
Tel (408) 721-5000
Fax (408) 730-0764
TWX (910) 339-9240

NEC

NEC America Inc
8 Old Sod Farm Road
Melville
NY 11747
Tel (516) 753-7000
Fax (516) 753-7041
Telex 144658 NECAMRMELV

OKI

OKI Semiconductor Inc
785 N Mary Ave
Sunnyvale
CA 94086
Tel (408) 720-1900
Fax (408) 720-1918

Panasonic

Panasonic Industrial Co
Semiconductor Sales Div
1616 McCandless Dr
Milpitas
CA 95035
Tel (408) 945-5672

Sanyo

Sanyo Semiconductor Corp
80 Commerce Dr
Allendale
NJ 07401
Tel (201) 825-8080
Fax (201) 825-0163
Telex 135138
SANYOSEMIALNJ

Seiko Epson

S-MOS Systems Inc
2460 N First St
San Jose
CA 95131-1002
Tel (408) 922-0200
Fax (408) 922-0238
Telex 176079 SMOS SNJUD

SGS–Thomson

SGS–Thomson Microelectronics
1000 E Bell Rd
Phoenix
AZ 85022
Tel (602) 867-6100
Fax (602) 867-6102

Sharp

Sharp Electronics Corp
Microelectronics Grp
5700 NW Pacific Rim Blvd
Suite 20
Camas
WA 98607
Tel (206) 834-2500
Fax (206) 834-8903
Telex 49608472

Texas Instruments

Texas Instruments Ltd
Microprocessor &
Microcontroller Products Div
Box 809066
Dallas
TX 75380
Tel (800) 232-3200

Toshiba

Toshiba America Electronic
Components Inc
9775 Toledo Way
Irvine
CA 92718
Tel (714) 455-2000
Fax (714) 859-3963

Index